Surface Application of Paper Chemicals

Surface Application of Paper Chemicals

Edited by

JAMES BRANDER
Head of Exploratory Research
Arjo-Wiggins Limited
Beaconsfield
Buckinghamshire, UK

and

IAN THORN
European Business Development Manager
Eka Chemicals
Weston-super-Mare
Somerset, UK

BLACKIE ACADEMIC & PROFESSIONAL
An Imprint of Chapman & Hall
London · Weinheim · New York · Tokyo · Melbourne · Madras

Published by Blackie Academic and Professional, an imprint of Chapman & Hall, 2–6 Boundary Row, London SE1 8HN

Chapman & Hall, 2–6 Boundary Row, London SE1 8HN, UK

Chapman & Hall GmbH, Pappelallee 3, 69469 Weinheim, Germany

Chapman & Hall USA, 115 Fifth Avenue, New York NY 10003, USA

Chapman & Hall Japan, ITP-Japan, Kyowa Building, 3F, 2-2-1 Hirakawacho, Chiyoda-ku, Tokyo 102, Japan

DA Book (Aust.) Pty Ltd, 648 Whitehorse Road, Mitcham 3132, Victoria, Australia

Chapman & Hall India, R. Seshadri, 32 Second Main Road, CIT East, Madras 600 035, India

First edition 1997

Typeset in 10/12pt Times by Cambrian Typesetters, Frimley, Surrey

ISBN 978-0-7514-0370-1

A catalogue record for this book is available from the British Library
Library of Congress Catalog Card Number: 97–73236

Printed on permanent acid-free text paper, manufactured in accordance with ANSI/NISO Z39.48-1992 (Permanence of Paper).

For our families

Contents

Contributors

J. Anderson	Eka Chemicals Ltd, 304 Worle Parkway, Summer Lane, Worle, Weston-super-Mare, Somerset BS22 0WA, UK
I. Attrup	Faxe Kalk, Frederiksholms Kanal 16, PO Box 2183, DK-1017 Copenhagen K, Denmark
G. Bacquet	Rhône Poulenc Recherches, Centre de Recherches d'Aubervilliers, 52 Rue de la Haie-Coq, F-93308 Aubervilliers, France
N.-O. Bergh	Cerestar Euro Centre Paper, Düsseldorfer Straße 191, D-4150 Krefeld 12, Germany
J. Brander	Arjo Wiggins Ltd, Research & Development, Butler's Court, Wattleton Road, Beaconsfield, Buckinghamshire HP9 1RT, UK
J.B. Clarke	Allied Colloids Ltd, PO Box 38, Low Moor, Bradford, Yorkshire BD12 0JZ, UK
D. Guest	Director of Technology, Masonite International Paper Co., John M. Coates Technical Center, 1955 Powis Road, West Chicago, IL 60185, USA
S. Hansen	Faxe Kalk, Frederiksholms Kanal 16, PO Box 2183, DK-1017, Copenhagen K, Denmark
J.-C. Isoard	Rhône Poulenc Recherches, Centre de Recherches d'Aubervilliers, 52 Rue de la Haie-Coq, F-93308 Aubervilliers, France
T. Kimpimäki	Paper Converting Institute, Tampere University of Technology, PO Box 541, SF-33101 Tampere, Finland
R. Rantanen	Valmet Corporation, PO Box 587, SF-40101 Jyväskylä, Finland
A.V. Savolainen	Paper Converting Institute, Tampere University of Technology, PO Box 541, SF-33101 Tampere, Finland
K. Stränger	Jagenberg Papiertechnik GmbH, Jagenbergerstraße 1, D-41468 Neuss, Germany

I. Thorn Eka Chemicals Ltd, 304 Parkway, Summer Lane, Worle, Weston-super-Mare, Somerset BS22 0WA, UK

A.S. Tindal Clariant UK Ltd, Calverley Lane, Horsforth, Leeds LS18 4RP, UK

M. Watson Arjo Wiggins Ltd, Research & Development, Butler's Court, Wattleton Road, Beaconsfield, Buckinghamshire HP9 1RT, UK

H. Weaver Renaissance Chemicals, Holly House, Brayton Lane, Brayton, Selby, Yorkshire YO8 9DZ, UK

Preface

With the exception of a slight hiccup during the height of the recent environmental movement (during the early 1990s), when for a year or two consumers were prepared to pay a price premium for lower quality recycled paper than for the virgin product, the inexorable improvement in the quality demanded of paper products continues.

This demand for quality covers not only the aesthetics of the product but also its performance. Moreover, it is becoming increasingly the case that papers designed for a particular use must, as it were incidentally, also perform well in alternative applications. An example is that of office and printing papers, which are expected to perform as well in copier machines as in all the various forms of impact and non-impact printers.

But even greater demands are made in other product areas, where board designed for dry foods can also be expected to protect moist and fatty materials and be made of 100% recycled fibre. The need to isolate foodstuffs from some of the contaminants that can affect recycled board is a serious challenge.

Thus, papermakers are constantly striving to meet a broadening spectrum of demands on their products; often while accepting declining quality of raw materials.

The product design philosophy that has arisen in response to this is increasingly to isolate the bulk of a paper from its uses: to engineer the needed performance characteristics into the paper surfaces while more or less ignoring what happens inside.

Thus surface application and treatment has assumed greater and greater importance, and a need for knowledge of the technologies and chemicals used has spread throughout all sectors of the industry.

A great deal of surface treatment is still performed in separate off-machine coating operations, using tolerably-well-understood technologies such as blade or roll applicators. However, the cost of such activities is obviously high, both in investment and in operational terms, and there is a constant search for ways to carry out larger proportions of the overall manufacture in a single operation.

Hence, interest in the paper industry and its suppliers in on-machine surface application technologies and chemicals is intense. Special attention is being given to finding binders, rheology modifiers and barrier polymers less susceptible to the film-splitting arising from the low-shear metering used in on-machine application systems; and in developing simply applied surface treatments adequately receptive to the dyes used in ink-jet inks.

Alongside these development efforts, on-machine surface treatment has an important role even when the final surface properties are provided off-machine. Paper and board containing a high proportion of recycled fibre simply can't be treated directly in high-shear metering systems such as blade coaters (which give the best results both for printing papers and for barrier films). It is simply too dirty; some form of pretreatment must be applied first. On-machine application of simple starch is ideal for this.

Happily, too, many of the surface treatments being applied, especially to food board, are replacing quite complex composite materials, including aluminium and polythene as well as paper. This means that the new, on-machine surface treated boards are actually easier to recycle than the materials they replace. This is important, as our supplies of virgin fibre are not going to increase as rapidly as our demand for paper and board in the coming decades.

This book describes some of the recent achievements in this field.

James Brander
Beaconsfield, Bucks
May, 1997

Introduction

J. BRANDER and I. THORN

Background

The extraordinary versatility of paper, its strength, flexibility, receptiveness to a variety of treatments, absorption of liquids and recyclability, and, ultimately, its destructibility means that it can continue to compete, despite the development of other materials, in many of its traditional uses. Accordingly, per capita paper consumption remains one of the most sensitive indicators of the state of a country's economy.

This combination of variety of uses to which paper is applied and the improving quality of its competing materials means that paper producers are continually needing to improve and diversify their own products. Because they are constrained both by the raw materials they can work with and by the technology available to them (modern paper machines still use essentially the same process as that invented by the Fourdrinier brothers two hundred years ago), they have found that making a relatively straightforward substrate and then modifying the paper surface with suitable chemicals can provide a cost-effective way of meeting the multitude of demands made by modern printing and packaging.

Paper made by a machine using a wet process is also intrinsically variable. This variability ranges from the distribution of fibres at the micron level to the distribution of their aggregates (flocs) at the millimetre level. This variation at the finer level is desirable since it provides a pore structure which gives paper one of its most important properties: the ability to absorb liquids. However, the variability in floc size and distribution is not desirable as it reduces smoothness, gives an uneven appearance to the sheet (called its formation), interferes with printing, gluing and other functional applications and reduces strength and opacity.

A good papermaker can do much to control this variability, enhancing those aspects that he or she knows improve the product, whilst suppressing the others. Unfortunately, there are limits to how far this process can go, and, ultimately, it may be necessary to use surface treatment simply to hide unwanted variability. Furthermore, the increased use of recycled fibre today adds another aspect to the variability which the papermaker has to contend with, because of the variations in waste-paper supply. It also greatly increases the contaminant load that the final product must be able to sustain without compromising its performance.

Other factors have also contributed to the rise in use of surface treatment; the conversion of many papermills in the USA to alkaline sizing resulted in some wet end size being substituted for surface size used in the size press and latterly film transfer presses to help solve problems of friction loss and size reversion in the paper: there is a recent demand for a universal office paper that can be used both for photocopying (requiring good toner adhesion) and for ink-jet printing where a combination of hydrophobicity and high print density is sought.

In the board industry, too, papermakers are facing demands for products possessing conflicting properties, such as resistance not only to water but also to oils and gases.

So, there are broadly two generic reasons for applying chemicals to the surface of paper: first, to provide the particular characteristics required by the variety of end uses to which paper products are subjected; and, second, to hide undesirable variation and contamination. Because modern developments are all tending in that direction, the subjects dealt with in this book have largely been orientated towards surface application that can be undertaken on the actual paper machine. Many of the chapters are relevant to both on- and off-machine application, but there is intended to be a bias towards still-developing technologies associated with the former type of application.

Technology of on-machine application

Dipping a material into a chemical bath and wiping off the excess is one of the most elementary means of surface treatment; it remains very widely practised in the paper industry today. The process is known as size pressing and has been used for a long time to apply chemicals to paper surfaces. In this case, the chemicals tend to be starches, though other treatments are also possible.

However, the size press is limited both in the range of chemicals it can apply successfully and in the speeds at which it can be operated without causing unacceptable production losses. Still, it is probably true to say that almost every paper machine in the world operating more slowly than 600 m min^{-1} has a size press. All of the chemicals discussed in this book can be applied at a size press although they may not attain optimum functionality by this means.

As papermaking speeds have increased beyond 600 m min^{-1} (many printing paper machines now operate at twice that speed) it has become important to find ways of improving on the size press; and as the chemicals used in surface treatment have become costlier and the reasons for adding them more demanding it has become essential to maximize the properties obtainable from them.

For a long time, the most effective process for achieving high-quality printing surfaces has been blade coating. This is a sophisticated operation about which a great deal is known, and so it was natural that the first attempts to improve on the size press operation concentrated on using blade coating on the paper machine. Chapter 1 (written by the foremost blade coater manufacturer in the world) describes these developments, their uses and their limitations. The fact that this approach to surface application has now been largely superseded by the technology described in Chapter 2 does not in any way detract from the achievements of on-machine blade coating which was, for some years, the only means of applying even simple surface chemical treatments to paper at speeds greater than the size press could reliably achieve. Indeed, although on-machine blade coating is presently eclipsed by the metered size presses described in Chapter 2, it may well receive a new lease of life as the limitations of the newer technology become appreciated. Thus, Chapter 2 describes the presently ascendant means of on-machine surface treatment by one of the engineers most closely associated with its development, whose organization pioneered the process and has developed it to its present sophistication. The chapter is a detailed scientific analysis of the process of metered size pressing; allowing, perhaps for the first time, predictions to be made as to how it might cope with applications outside its present range.

However, while this technology, known also as the film press, has made possible many advances in surface treatment and has, in principle, opened up the possibility of producing fully coated magazine printing papers on a single machine compared with the two which the product currently needs, it remains tantalizingly deficient in some key quality aspects. The reasons for this, though understood, have yet to be resolved.

The reason for this quality deficiency, however, is the characteristic that makes possible another important achievement of this technology: providing a way of hiding many of the structural defects and contaminants in recycled papers. Without the metered size press, recycled fibre would need to be much cleaner before it could be used in many of the present products in which it is included: for example, coating base paper, packaging board and many printing grades. Cleaning of the recycled fibre to achieve the same result would greatly increase costs and further diminish the environmental benefits that recycling brings.

Metered size press technology is developing very fast, and many of the chemicals described in the following seven chapters are evolving as well; many papermakers are attempting to close the quality gap between chemical application by on-machine metered size press and off-machine blade coater.

Chemicals for surface application

Perhaps the chemical most likely to spring to mind when discussing surface treatment is starch; of the more than one million tonnes of starch supplied to the European paper industry only a small fraction (16%) is added at the wet end of the machine. The rest is either sprayed, surface applied or used as part of the binding element of coating mixes. Although native starch is sometimes used, chemical modification is often necessary to attain more sophisticated properties. Chapter 4 details how such properties as film formation, water retention, binding strength, etc., are improved by a variety of changes to the natural chemistry of starch. Starch degradation has the effect of reducing the viscosity of the product in solution and thus makes it more easy to handle. This is achieved by a number of treatments, these being: the use of acid, enzymes, or a combination of heat and oxidants. In addition to reducing starch to lower molecular weight fragments, other chemical treatments are used to endow other properties on the material; for instance, the use of hypochlorite, chlorohydrins or ethylene oxide can introduce such properties as reduced retrogradation, improved film formation and higher water retention, respectively. The importance of some of these functionalities is discussed in Chapters 3 and 7 (film formation) and Chapter 5 (water retention).

Starch has traditionally been applied at the size press and latterly at the metered size press, but not so routinely with a short dwell-time blade coater (this, too, may gain more acceptance in the future). When starch is used to prevent curl it is added to the non-coated side of the sheet often by a technique called liquid application system (LAS).

Starch is not always the product of choice and is often wholly or partly replaced as a binder in some applications. In this case, synthetic latices are employed. Latices are emulsion polymers, usually produced from vinylic monomers such as styrene, butadiene, acrylic acids or a variety of acrylates. Combinations of these and other monomers are polymerized by radical initiation within micelles of surfactants so that the resulting polymer is in the form of an emulsion or latex. The reason for producing the polymer in this form is that it is more easy to handle and has a rheology more compatible than the corresponding solution polymerization product.

The choice of monomers is determined by the properties required from the final coating and the restrictions of the surface application apparatus. Some monomers when polymerized alone will give a soft, more flexible film and others a harder, more brittle film. Through judicious choice, two or more monomers can be copolymerized to give the final film properties required by the final treatment of the surface.

Film formation is not the only function of the latex, since like the pigment dispersants detailed in Chapter 5 the products must also wet the surface of the pigments used in the coating mix, to ensure that good binding takes place. Chapter 3 also discusses the effects of latices on the surface energy of the paper; important because this property has a direct bearing on such printing properties as ink refusal and print-dot size, as well as affecting the edge binding of papers when hydrophilic glues are applied.

There has been a growing demand for papers of higher smoothness, printability, opacity and brightness. Although the use of starch or latex binder improves some of these properties, pigments, commonly high-surface-area clays or calcium carbonate, must be used to improve the rest. The mixture of pigment and binder, known as the coating colour, has been traditionally applied to the surface of paper by the use of a blade coater. Chapter 6, however, offers an alternative technique for applying a pigment mixture onto the surface of paper, on-line. With the development of the metered size press, the possibility has been presented to apply a pigment and, typically, starch only onto the paper surface; this technique has been given the name 'pigmentizing' to differentiate it from the more common off-machine coating.

Pigmentizing is a cheaper alternative to coating when only a light weight (2–4 g m^{-2}) is required. The type of pigment used in this technique depends upon the purpose of pigmentizing; if pigmentizing is undertaken to produce a pre-coat, then coarsely distributed pigments are used such as marble or coarse clay; whilst for general paper property improvement, such as porosity, high-surface-area pigments are used.

As coating speeds have increased it has become even more important to exert control over the rheology of the coating colour since this has a profound effect on coat weight uniformity, pigment binder distribution and some important surface properties. The manner in which rheology is modified and maintained is discussed in Chapter 5 in some detail, although the significance of this parameter is also mentioned in Chapters 3, 4 and 7. Even though these products are sometimes referred to as 'thickeners' this term undervalues the complexity of their operation and their chemistry; the structure of these materials, for example, can vary between an essentially linear, water soluble polymer to a cross-linked gel-like state which resists the stretching that occurs when shear is applied. In this way shear stability is built into the coating mix, and less variation is observed at different machine speeds and coating processes. The viscosity modifier also acts in concert with some of the binders described in Chapter 3 to form three-dimensional structures which again add to the shear stability of the coating colour. The other type of viscosity modifier covered by this chapter is dispersants, which, although varying in chemical nature, wet the pigment powder, de-agglomerate pigment clusters and stabilize the dispersed particles to prevent future agglomeration.

The growth of ink-jet printing has seen a hasty search by equipment manufacturers, ink producers, papermakers and chemical suppliers (sometimes in concert) to unearth the factors which determine good printability.

However, even before ink-jet printing became so popular, sizing was often carried out at the surface of the paper. Addition of chemicals at the surface rather than at the wet end of the paper machine has been an attractive proposition to some papermakers for some time; not least of their concerns is the increasing closure of the whitewater system and the corresponding increase in contamination. If proof were needed of this move to surface application one need only read the statistics in Chapter 4. Chapter 7 explores the types of chemicals that are added at the size or metered size press and their effect on the surface properties of the paper with special regard to sizing and ink-jet printability.

Because even bleached pulp is intrinsically yellow, blue dye or pigment has often been added to compensate. This, however, sometimes results in the paper obtaining a grey hue, so an alternative practice is to add fluorescent brightening agents (FBAs) which absorb ultraviolet light and re-emit it in the range of 435–445 nm. Though FBAs can be added at either the wet end or the surface, Chapter 8 deals solely with the latter. Addition of FBAs to the surface ensures that 100% of the product is retained; some is also added at the wet end to mask imperfections in the surface application. Chapter 8 also examines the causes of 'greening' (an effect of aggregation of the FBA molecules) and how it can be avoided; in addition the effects of recycling, quenching and temperature are covered.

As well as making the paper white, surface application can also be used to colour the paper (Chapter 9). Dyes can be added to the surface by spraying or by addition to the size press or coater. In much the same way as with FBAs, these processes are often done in addition to stock treatment at the wet end. The difference between stock dyeing and that at the size press is that the dye reacts with the starch as the substrate rather than the cellulose fibre. This, however, means that any imperfections in the coating process can be clearly seen in contrast to a white base sheet. Colours achieved by size press dyeing may differ from that expected from the shade card since these apply to stock dyeing. Both advantages and disadvantages are covered in this chapter.

Paper products and surface treatment

After discussion of the role of chemicals, both alone and in blends, the following two chapters focus on particular product areas in which surface treatment has assumed particular importance: papers for modern non-impact printing (ink-jet and fused-toner); and papers and boards with the barrier properties required for use in food and liquid packaging. Both are areas of intense interest.

It is the accepted view that ink-jet printing provides the only viable means of locally rendering into hard copy the colours which are now readily available on transient imaging systems. Thus, much development work in the paper industry is focusing on this area, attempting to produce papers with the ability to reproduce colour images without the costs in processing and raw materials associated with the present coated papers.

Until recently, difficulty had arisen because the inks had to be almost pure water in order not to plug the very fine nozzles used to dispense them. Except for some very fine black pigments used in most of the recently developed printers, all the inks have been based on dyes. The problem faced by those developing paper products has been to fix the dyes at, or very close to, the surface while absorbing the water into the bulk of the paper. Generally speaking, it has been possible to achieve one of these ends satisfactorily, but not both together on uncoated paper.

What is needed is a very thin, very competent film with a strong affinity for dyes above a very absorbent paper substrate. In addition to this, paper made in this exacting way must also perform satisfactorily in the major other office use today: laser printing. The papermaker's attempts to rationalize these competing demands are described in Chapter 10.

Laminating and extruding are processes that have been in use for many years to provide the barrier properties required by board used in packaging. In the former, adhesives are used to secure thin polyolefin films or aluminium foil onto the board, while in the latter the polyolefin film is extruded directly onto the board and then chilled into place. However, both processes are difficult and expensive, often involving multiple extrusion or laminating heads on the same machine. Achieving a correct match between the properties of the various materials and the board so as to avoid curl was always very difficult, and the restrictions on the use of organic-solvent-based adhesive has made this even worse. So, Chapter 11 describes the polymers, that properly applied, can replace many of these laminates, allowing satisfactory barrier properties to be achieved by the technically much simpler process of coating.

Hitherto, adequate barrier performance has only been achieved by off-machine coating, using either blades or various roll applicators. However, work to develop these polymers for use in metered size presses is intense, opening up the possibility of creating a complex packaging laminate on a single paper machine, often based on 100% recycled raw material.

The final chapter (Chapter 12) considers the whole subject of surface application of paper chemicals from the perspective of recycling the waste products after use. Issues of 'white pitch', 'stickies' and excessive amounts of pigment in whitewater circuits are the well-known disadvantages of recycling coated papers. Overcoming some of these chemical effects can create other difficulties, which this chapter explores.

The advent of the laws in support of the EU packaging directive, and the levying of landfill taxes in several EU states, means that this whole issue is taking on increased importance. It is especially relevant to the polymer dispersion coatings discussed in Chapter 11, which, by and large, are more easily recycled than many of the composites they replace.

Since this book was originally proposed there has been a new development in non-impact printing: the digital press. This represents another route to the goal of rapid realization of high-quality colour hard-copy, though designed for longer runs than ink-jet. The demands made on paper by these processes (two are competing) are entirely different from those of ink-jet and laser and represent a new spectrum of needs for surface application of chemicals. Clearly, any future study of this subject will have to include these.

1 On-machine surface treatment of paper and board with the blade coater

K. STRÄNGER

1.1 The origins of on-machine surface treatment

Literature provides only vague references as to when paper and board first started to be surface-finished on-line. We can only be certain of the real industrial breakthrough, which came with the introduction of the roll coater, based on an invention by Peter Massey. The first commercial installation was commissioned in 1933 at the Consolidated Water Power and Paper Company in Wisconsin Rapids, Wisconsin.

In contrast, literature offers plenty of references to the blade coater's first applications, the first printed patent specifications dating back to shortly after the turn of the century. The actual invention of the blade-coating system is most frequently attributed to Arthur Ronald Trist, from London, who was awarded US patent number 2.368.176 on 30 January 1945 for his trailing blade puddle type coater (Figure 1.1). This blade coater was first commissioned as an off-machine coater in 1955 and installed on-line in 1958 at the Gulf States Paper Corporation, in Demopolis, Alabama, for the surface treatment of food board.

Two other on-machine blade coaters dating from 1958 are worthy of mention, since they were the first to be equipped with the inverted blade coater (flooded nip coater) and an applicator roll (Figure 1.2). The first was installed by the Gilman Paper Company, in St Mary's, Georgia, for surface treating food board. The second was Europe's very first on-line blade coater, commissioned in 1958 by Enso-Gutzeit OY in Kaukopää, Finland, also for surface treating food board.

The inverted blade principle, based on US patents 2.746.877 and 3.097.107 awarded to Rush and R.W. Martinek, respectively, is that found at the heart of all today's blade-coating systems, although the application method and blade-system function can vary.

1.2 The blade-coating system today

On-line blade coaters do not essentially differ from their off-line counterparts, although some additional equipment may be necessary on

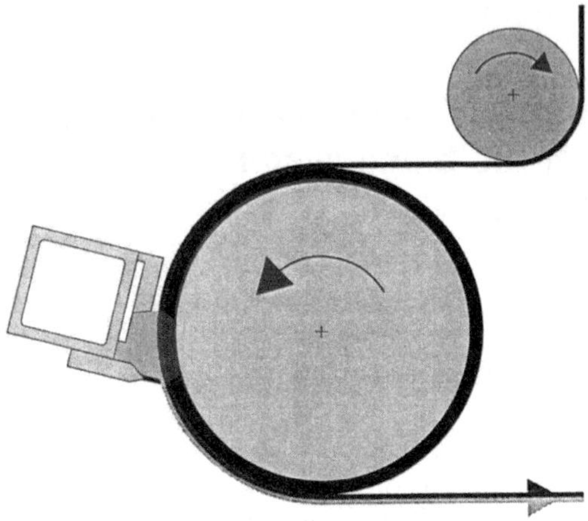

Figure 1.1 Trailing blade puddle type coater.

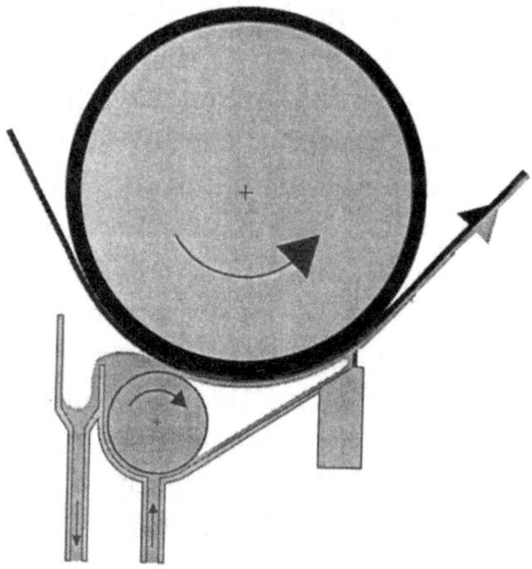

Figure 1.2 Inverted blade coater.

the on-machine coater to safeguard the efficiency of the production line.

Depending on the working-speed, substrate, desired dry coat weight and coating media, the following five application systems can be said to predominate (Figure 1.3):

- the conventional roll-application system;
- the roll-application system with infeed profile;
- the short-dwell coater;
- fountain applicators;
- the applicator nozzle with premetering unit.

In the low-to-medium speed range (up to approximately 700 m min^{-1}), which is the normal speed range for board coating or speciality paper production, conventional roll-application systems are most frequently used. For operation at higher speeds, these systems are usually equipped with an infeed profile or applicator nozzle.

The technique for adjusting the blade load and blade angle in order to meter and spread the coating medium varies from supplier to supplier. Ultramodern systems can meter in both the rigid and bent-blade mode and with the metering rod as well (Figure 1.4). To secure continuous production, the machine design must also allow for rapid changeover from one system to another.

For maximum uptime with minimum broke, the on-line coating head should be equipped with the following functions:

- pneumatic blade clamping with water jets to clean the clamp so that the blade may be exchanged in the shortest possible time;

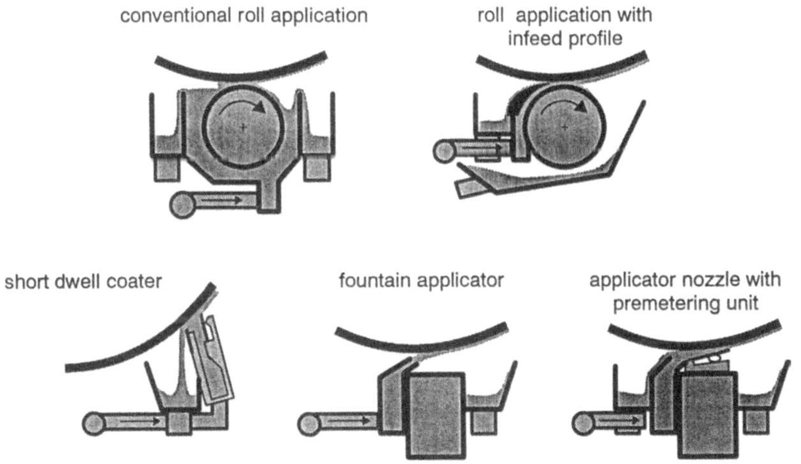

Figure 1.3 Applications systems for blade coater.

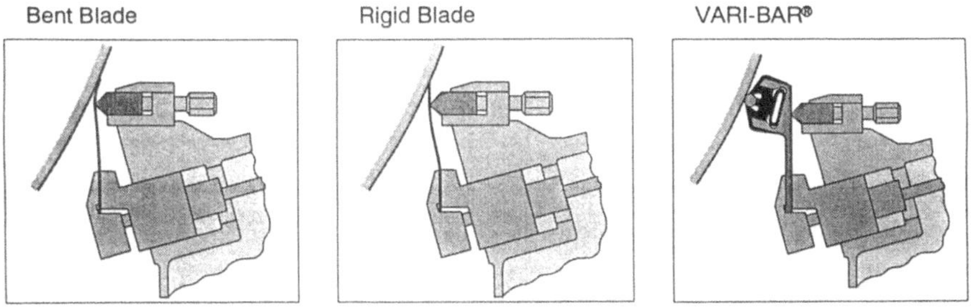

Figure 1.4 Metering systems for blade coater.

- reproducible setting of the following parameters by means of motor-driven stops and digital displays:
  - gap between applicator roll and backing roll;
  - gap between blade beam and backing roll;
  - blade angle;
  - blade load;

  such reproducible setting of the major parameters makes for swift machine make-ready following a product changeover and facilitates automatic coat-weight regulation via the quality control system through blade-angle and blade-load adjustment;
- web moistening by means of water sprays arranged across the working width to prevent the blade running dry during engagement and retraction of the application and metering systems;
- cleaning doctors on the backing roll to protect its surface during production from coating colour which has seeped through the substrate and from other impurities, for example, fibres;
- cleaning apparatus for rapid cleaning of the application system and backing roll following a web break so that operation can be resumed quickly;
- web-severing units ahead of and following each coating head to prevent the web from wrapping itself around the backing roll after a web break.

1.3 Advantages and disadvantages of on-machine surface treatment

There is no definitive answer to this question. Individual manufacturers must decide for themselves by looking at the present production conditions and those planned for the future, by drawing on their own experience in the manufacture of coated products, and basing their conclusions on an analysis of the market, the anticipated market development and long-term product development plans.

Generally valid arguments in favour of the on-line coater do exist, however:

- lower investment is needed in building and machinery;
- lower costs are involved for materials, energy and personnel;
- lower costs are involved for transport and storage between the paper or board machine and subsequent converting machinery, for example, winders and sheeters.

If these cost advantages are to have a really positive effect on a company's output, a number of preconditions must be fulfilled. These include:

- perfect functioning of the paper or board machines and use of a substrate which is free from holes, tears, fibre agglomerates and other mechanical defects which could cause a web break in the coater;

Table 1.1 Advantages and disadvantages of on-line and off-line coaters

Differing characteristics	Evaluation	
	Off-machine	On-machine
Additional re-reeler between paper machine and coater	−	+
Intermediate reel storage	−	+
Additional coater unwind and flying splice	−	+
Additional coater rewind	−	+
Space requirements (building)	−	+
Personnel requirements for material handling and operation	−	+
Energy costs	−	+
Flexibility (scheduled downtimes, grade changes)	+	−
Production losses as a result of quality defects	+	−
Ability to adapt to quality fluctuations in the substrate	+	−
Web threading in coater	+	−
Stable quality	+	−
Temporary changeover to multiple coats	+	−
Adaptation of machine speed to the coater requirements	+	−
Number of web breaks	+	−

- infrequent product changes, that is, the individual order must be for as large a quantity as possible;
- sufficient available space in the paper or board machine for installation of the coater;
- steady web guidance with free web runs within the coater kept to a minimum.

The pros and cons must be weighed up carefully before a decision is made. With on-line coaters, mechanical defects in the substrate, web-run problems in the coater or too frequent product changes can seriously affect the machine uptime and diminish the performance of the entire production line. Table 1.1 lists the main points of comparison. These only serve as a guideline, however, and a definitive assessment can be made only with reference to the individual mill or production line.

1.4 Paper and board grades utilizing on-machine surface treatment

At the beginning of the 1960s, the US and European markets started to develop in different directions. In the USA lightweight coated (LWC) grades and other types of mass-produced printing papers were predominantly blade-coated on-line. In Europe, this task was usually performed off-line. This is explained by the fact that the US market for these paper grades was much bigger than in Europe and therefore favourable to continuous on-line production without interruptions for product changes.

From the beginning, both in Europe and the USA, board was usually blade-coated on-line (for example, the first applications in 1958, described in Section 1.1). This is mainly on account of the lower speeds at which board substrates are produced and their much lower propensity to web breaks.

Since the mid-1980s, on-line coating has also become established in Europe for LWC and other mass-produced printing papers and also for speciality papers, such as those used for labels and flexible packaging.

Ultramodern boardmaking technology no longer just means coating folding boxboard on-line. The on-line coaters of today are sophisticated enough to output graphic board and superior boxboard grades of the highest gloss and smoothness, used, for example, to pack cosmetics.

The increasing trend towards coating on the paper or board machine was and continues to be encouraged by the increasing reliability of paper and board machines, by the higher degree of automation, especially of product changes, by the scope for combining different application systems within a plant and, no less significantly, by constantly improved calendering technology for pretreatment and post-treatment on-line.

The coating quality which can be attained on a given substrate, particularly in terms of opacity, gloss, smoothness and printability, is not only a question of suitable coating colour formulations but also of the right choice and number of coating heads and dryers. These must be selected carefully, bearing in mind the overall coating concept.

1.5 Designing the surface treatment line

Besides reliably functioning coating heads, the web run and facilities for web threading at production speed decisively influence the efficiency of the coating line.

Special attention should also be paid to the provision of sufficient space for operation and maintenance, especially at coating heads and dryers, and of adequate facilities for disposing of broke, such as pulpers and conveyor belts.

1.5.1 Web run

The primary objective when designing the layout of an on-machine coater is to find the best way of guiding the web through the coating line without web-wandering, creasing or web breaks. This also applies to the acceleration and deceleration phase during a product change and to load changes, caused, for example, by engagement and retraction of the application and metering system.

The following factors are crucial to a good web run:

- short web runs between fixed points;
- selection of the right roll diameters, particularly where the angle of wrap is large;
- suitable distances between rolls and sensible use of spreader rolls;
- stable fixed points connected to electronic load cell and regulating equipment for drive control, and drives for all guide and spreader rolls.

Generally speaking, we should strive for short, free, web runs and the shortest possible distances between fixed points and seek to avoid large angles of wrap on small-diameter rolls, particularly at points where the coat is only partially dried.

1.5.2 Web-threading

Unlike the off-machine coater, the on-machine coater requires web-threading at working speed. To this end, a web tail is made in the cylinder group ahead of the coater and is pulled through the coating section using two or (better) three parallel threading ropes. The ropes should run over free-running rope pulleys and be driven by separate motors so that they can move at a different speed from the web if required.

The distance between the rope pulleys should be such as to exclude the possibility of the ropes wandering apart, which can cause the web tail to slip between the ropes. Web-tail slippage can lead to web-tail breakage and a failed attempt at web-threading.

The degree of automation available for threading the web tail into the rope nip at the start of the coater varies from supplier to supplier.

Particularly on lines with several coating heads, excessively long rope runs should be avoided and the rope run divided into sections, each with its own rope drive and tensioner.

Hence, the efficiency of the entire production line is not just a question of the right coating concept, but also of a well-designed and properly functioning rope run.

1.6 Examples

The layout of today's standard coating lines is determined both by the end product and related quality requirements and by the manufacturer's own experience with existing coating plants. The coating concepts discussed in the following examples have been put into commercial operation, and each is an example of a tailor-made solution to a particular coating problem. Continuing advances in coating, drying and calendering technology and improved raw materials for substrates and coating colours will naturally inspire new ideas.

1.6.1 Folding boxboard

A significant proportion of recycled paper goes into the production of folding boxboard. Excellent optical properties and good printability are demanded of even 100% recycled board. Thus three coating heads are usually used today for coating the side to be printed. This task is performed either by three blade coaters, which can be alternatively run with the metering rod, or by a system combination of blade (or rod) plus airknife plus blade. The latter arrangement is usually chosen when the aim is perfect coverage of the board surface. An example of such a combination is shown in Figure 1.5. The precoat applied by the first coating head evens out absorption capacity and creates a smooth surface. Coating head 3 is an airknife system, which serves to improve the optical properties, particularly whiteness and opacity. The fourth coating head works in the bent-blade mode to produce the superior smoothness essential for maximum print gloss, without additionally having to calender. Coating head 2 carries out reverse side treatment.

With this coating concept, the desired quality in terms of smoothness and gloss is produced entirely by the blade coater without finishing by calenders or other surface treatment equipment. This places considerable demands on the bent-blade function and the coating colour formulation.

1.6.2 Graphic board

Graphic board is used for the covers of catalogues, annual reports and menus, and for invitations, greetings cards, calendars and posters, etc. Thus high printability is required on both sides of the board, and the chief objective of the coating procedure is to avoid discrepancies between each coated side of the product.

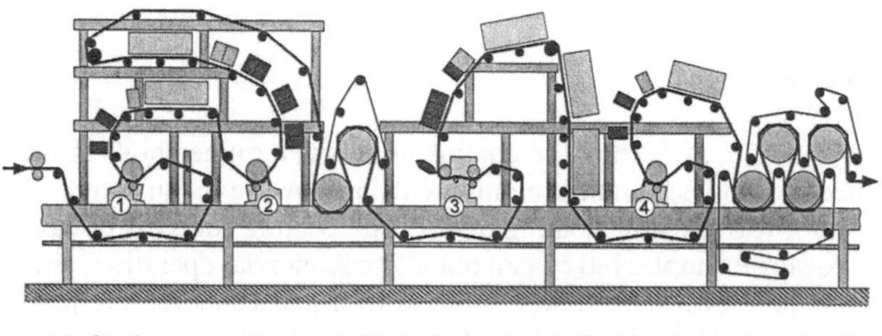

Figure 1.5 On-line coater for recycled board, showing coating heads 1–4 described in text.

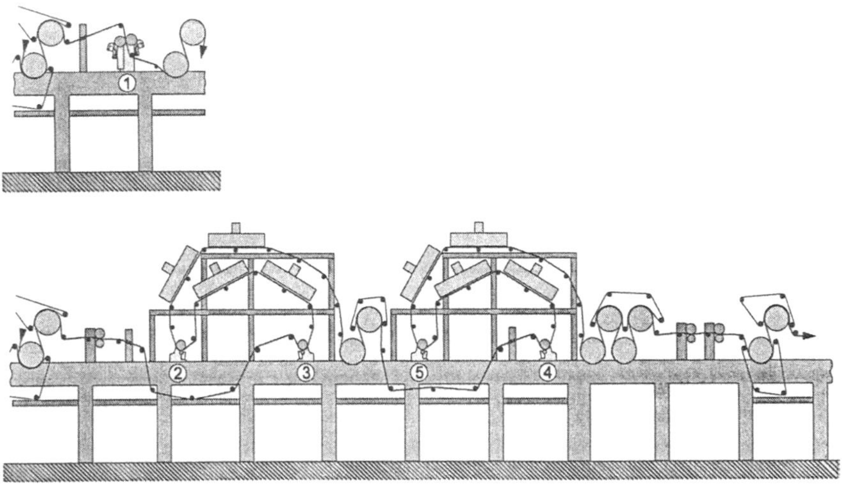

Figure 1.6 On-line coater for graphic board, showing coating heads 1–3 and blade coaters 4 and 5 described in text.

An example of a coating plant for graphic board is shown in Figure 1.6. Coating head 1 is a film press which treats the substrate on both sides with starch or a pigment precoat. This makes for uniform absorption properties and prevents any subsequent pigment coats from penetrating excessively into the web. The precoat is then applied on each side by coating heads 2 and 3, followed by blade coaters 4 and 5, which apply the top coat in bent-blade mode. Since a high degree of smoothness is already achieved by the bent blade, the following calenders, which also serve to add gloss, need only to be run at a low pressure, thus keeping the stiffness and bulk of the finished board product intact.

1.6.3 Low-coat-weight papers

Up to the present day, the blade system has been the preferred system for applying low coat weights to wood-containing papers (for example, in LWC production), since it achieves with ease the degree of smoothness required for subsequent rotogravure printing. The concept shown in Figure 1.7 succeeds in satisfying the twofold requirement of short web runs between fixed points with optimum space for operation and maintenance.

Coating head 1 coats the top side, and coating head 2 the wire side. An airturn is installed after the radiators on the second coating head, which turns the web without contact on the freshly coated side, thereby creating a highly economical web run. Following each coating head, drying is mainly performed by cylinders so that space-consuming airdryers, which impose longer free web runs, can be dispensed with.

Figure 1.8 shows a similar concept to that illustrated in Figure 1.7, with the same output. Again, the top side (1) and then the wire side (2) is coated. Here the cylinder group is shorter than that shown before, this being compensated for by airdryers following the infra-red section.

This system combination of infra-red, convection and contact drying creates greater flexibility and more scope for using the dryer setting to attain the required coat quality. The free web runs between fixed points become longer, however.

1.6.4 On-machine coated wood-free papers

In Figure 1.9 a coating plant for the manufacture of wood-free coated papers in the medium quality range is shown. Coating head 1 is a film-press system which applies a precoat of 2–5 g m^{-2} to each side. Coating heads 2 and 3 are run in the bent-blade mode to produce a two-sided top coat. In this case, post-treatment does not take place on-line; the coated paper is given its matt or glossy finish in a separate process.

1.6.5 Speciality papers

In Figure 1.10 an on-machine coater for manufacturing speciality papers coated on one side is shown, such as those used for flexible packaging, gift wrapping, labels, etc. The paper web is first coated with the film press on the wire side, reverse-side treatment with starch being performed simultaneously to give the one-side-coated finished product sufficient flatlay. Coating head 2 applies the top coat in the bent-blade mode, and further reverse-side treatment is carried out by coating head 3, a roll moistener. The speciality papers are not only coated on-line but are also surface finished by a double calender to add the smoothness and gloss required by the end product.

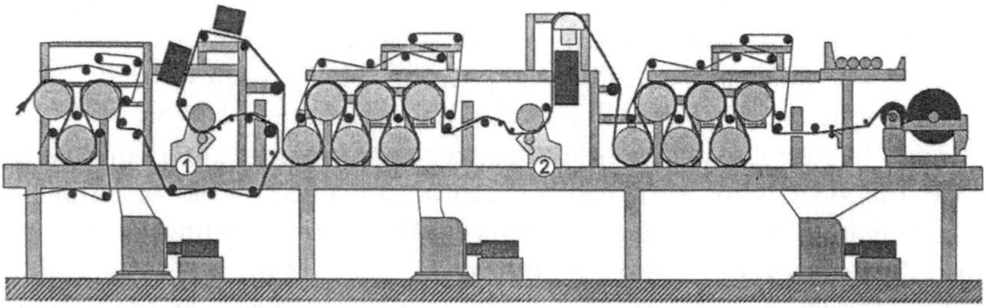

Figure 1.7 On-line coater for low weight coat papers, with infra-red radiators and drying cylinders only, showing coating heads 1 and 2 described in text.

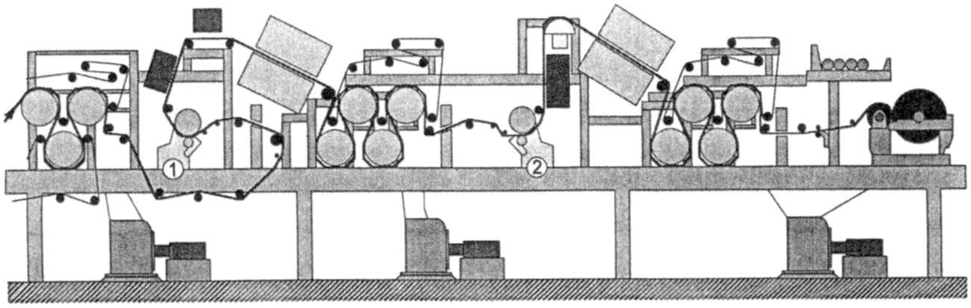

Figure 1.8 On-line coater for low weight coat paper, with infra-red radiators, airdryer and drying cylinders, showing coating heads 1 and 2 referred to in text.

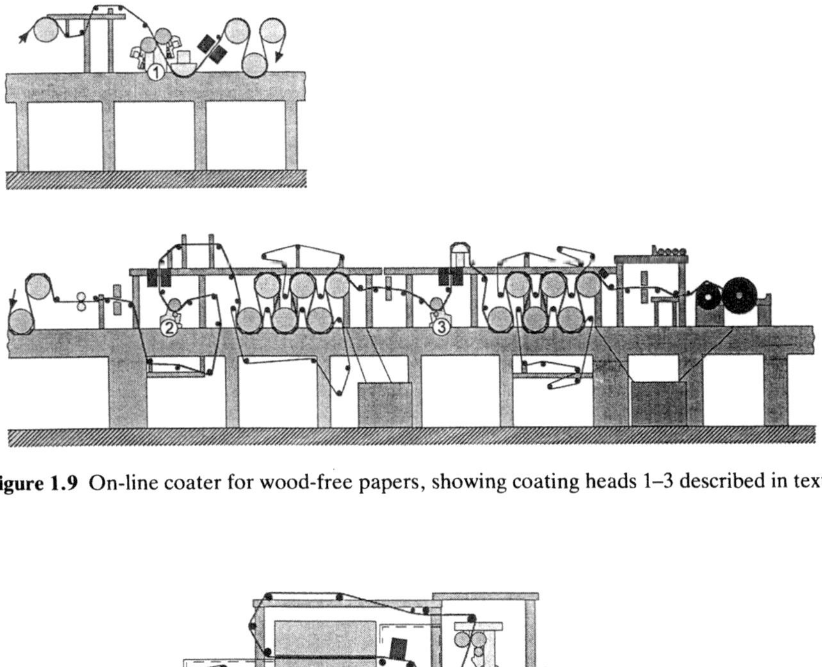

Figure 1.9 On-line coater for wood-free papers, showing coating heads 1–3 described in text.

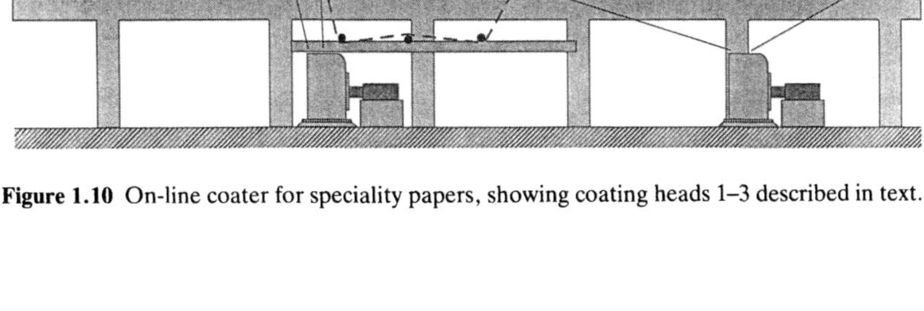

Figure 1.10 On-line coater for speciality papers, showing coating heads 1–3 described in text.

1.6.6 Conclusions

Today, the finishing of numerous end products is entirely realized within the paper or board machine using the blade-coating system. In conjunction with other coating systems and modern calendering systems nearly all the product requirements in terms of whiteness, coverage, gloss and smoothness can be satisfied on-line.

Nonetheless, in the field of wood-containing papers with low coat weights, even those with increasing recycled content, there are more and more instances of the blade coater being successfully replaced by the film press, especially in the production of papers destined for offset printing. The signs are that even LWC production, the classic domain of the on-line blade coater, will benefit from film-press technology.

2 On-machine surface treatment of paper and board with the film press

R. RANTANEN

2.1 Introduction

2.1.1 Review of surface sizing equipment up to the 1980s

Size press technology has undergone a revolutionary change over the past ten years. The driving force behind this was the speed increase of fine paper machines which the conventional pond size press was incapable of matching. New generation film presses were born, with no speed limits and with the flexibility to produce different kinds of surface sizing treatments, until finally the new process has become a challenger to the blade coating process itself. The competing blade surface sizing technology lost its chance because of low efficiency and inadequate paper quality. The gate-roll film press, even though having good efficiency, also has quality limitations as well as too low a speed limit in coating and in surface sizing, especially with high size amounts.

This chapter deals with experiences the writer has gained from working in 1965 as a trainee student at a six-roll film press producing art paper; from the first experiments carried out with a film press having a blade coating type film applicator at Enso-Gutzeit's Research Centre in 1975; and, since 1986, from developing the SymSizer film press and making comparisons with the gate-roll film press at high speeds on a pilot scale. The text does not contain a complete literature review but includes some literature findings which help in the basic understanding of film press technology. Quite little attention has been paid to numerous papers presenting results in producing different paper grades with film presses, because, today, there is a rapidly increasing supply of this kind of material. For example, multilayered coating technology, including the wet-on-wet process which has gained big interest, has not been discussed.

In the history of the new generation film press, which began with the development of the equipment, continued with technology adaptation by paper mills coupled with paper quality development work by paper mills and chemical manufacturers, a new era has now begun when scientific research teams have already published the first excellent theoretical studies of the processes. For papermakers, this development guarantees future paper quality competitiveness when using this low-cost, high-efficiency technology in production.

2.1.2 Metered size press technology

Size presses. Up until the 1980s, the conventional two-roll size press was standard equipment used for fine paper and board surface sizing and low solids pigmenting. Attempts at high-quality pigment coating for magazine paper grades led to a wide variety of film transfer roll coaters [1]. This development started on production scale in 1933 and was based on the utilization of a large number of rolls in premetering. This development gave way in the 1960s to the well-known gate-roll coater. Even though the gate roll enjoyed great success in surface sizing of fine papers and precoating of various paper grades, the invention of the blade coater prevented stronger interest in this field of technology.

The gate roll, however, gained a strong foothold later on in the Far East, especially in Japan, where during the 1980s it became the standard size press for all surface sizing and precoating as well as many single coating applications. During the same period, only a few new gate rolls were installed in the United States and even fewer in Europe.

Two-sided surface treatment based on blade technology. Based on blade technology, a solution for simultaneous two-sided surface treatment, called the Billblade coater, was invented late in the 1960s by AB Inventing of Sweden. This equipment consists of a puddle-type blade on one side of the sheet but also a puddle between the sheet and the backing roll. This blade/roll principle and its later modifications gained a good approval both in fine paper surface sizing and especially in treating one side with coating and the other with surface size [2]. Another example based on similar technology is the Twostream coater [3], where the web runs upwards to minimize the space requirement and to give good visibility of the outcoming sheet.

Other surface sizing technologies. The liquid application system (LAS) by BTG has a wide use in decurling. This process is also used for different types of treatment in speciality paper production. Its suitability for surface sizing and pigmenting of wood-containing grades has been studied, for example, by Pratt and Muenchow [4] with promising results, but the process never gained any importance in standard surface sizing or pigmenting applications.

Also in the 1980s, some short-dwell blade coaters were installed to run high-speed fine paper surface sizing [5].

2.2 Quality and runnability aspects of different surface sizing processes

2.2.1 Two-roll size press

To improve printing and strength properties, the conventional size press holds no great qualitative disadvantage in fine paper surface sizing applications [6]. Bone-dry size amount is easy to control through size solids content to a wide extent. In the pond and/or nip, size deeply penetrates the sheet, which has normally been considered beneficial for paper quality. The two-roll size press is also reliable: it requires little maintenance, and paper quality is constant. At lower speeds, the runnability is reasonably good.

However, at higher speeds (over 800 m min^{-1}), the size press is the bottleneck for the fine papermaking process because of an unacceptably high number of web breaks. Even though high-speed problems relating to pond splashing were quite eliminated through, for example, the Sym-Roll size press, it was impossible to reduce the number of breaks at high speeds.

The break reasons can be explained by at least three different mechanisms.

- Hydrodynamic forces in the pond rupture weak spots in the web.
- Holes and impurities in the paper web entering the size press can collect excessive amounts of size from the pond, causing sticking to the top roll, thus tearing the web.
- The paper gets so wet in the size press that its strength is too low in the afterdryer section.

In Europe, the two-roll size press has been commonly used as a simple precoating technique for conventionally coated wood-free papers. Size press pigment addition is also used to improve the printing characteristics of normally uncoated papers to produce a form of matt offset printing paper, and pigment is added at the size press in some mills to reduce porosity and generally to increase the pigment content of conventional uncoated fine papers. Heavier basis weight fine papers have been coated up to 8 g m^{-2} per side with maximum solids content 50%, but the speed has been very low, about 300 m min^{-1} [7]. For low solids pigmenting from a 20% solids content mix, the maximum speed is about 600 m min^{-1} giving a coat weight of 4 g m^{-2} per side.

2.2.2 Gate-roll size press/coater

Quality wise, the gate-roll size press is good for applications where very low size amounts are needed, such as in newsprint surface sizing. The major disadvantage is the impossibility to achieve sufficient size penetration for many applications. Size metering is based on hydrodynamic forces and so only by high viscosities is it possible to achieve thick size films on the applicator rolls. Such a size film lies on the sheet surface failing to penetrate the sheet.

With respect to web breaks in surface sizing, the runnability is good. Especially in coating, the major runnability problem at high speeds is misting after the nip between transfer and applicator rolls [8].

2.2.3 *Blade surface sizing*

In blade coater surface sizing, the size solids have to be twice as high as in pond surface sizing to achieve the same bone-dry size amount. The blade mainly fills up the surface roughness of the sheet, and so the applied wet size amount is very low. The quality of the paper surface is quite similar to that of the gate-roll surface sized sheet.

With blade/roll process, size staying on the surface has a benefit in a better surface strength than is achieved with the same bone-dry size amount with a pond size press. The other blade/roll surface sizing benefits are the much lower water content to evaporate in the afterdryer section, and, in board production, 10% lower density compared with the pond size press [2].

In spite of blade/roll type units having a different type of metering process on different sides of the web, there are no reported problems in sheet flatness properties.

Two-sided coating blade/roll type coaters are suitable up to speeds of 700 m min^{-1}, after which speeds the solids content of the coating colour on the roll side has to be reduced too much to achieve a reasonably good coating quality. Blade coaters installed for surface sizing achieved the speed targets but, quality wise, sheet flatness was impossible to guarantee.

Both blade and blade/roll type surface sizing processes never gained any significant success in surface sizing on big production lines. One reason was that the paper quality was different or 'worse' than that from the two-roll size press. The other reason was web breaks because the blade process is always very sensitive to base-sheet defects and impurities.

2.3 New generation film presses

The gate-roll size press had shown its excellent runnability in production but there was a great need to get away from the tough relationship between premetered film thickness and viscosity combined in this process, the complexity, and the high price of the equipment as well as relatively high maintenance costs.

The new type of film presses based on short-dwell coating heads came into production in the mid-1980s. The metering elements were bent-blade or grooved rod. Patents and patent applications from that time are as follows:

- patented new coating head intended for bent-blade film premetering and blade coating, a modified slot orifice type applicator [9];
- a patent application for a web upwards design, premetering with grooved or smooth gate-roll or puddle-type blade [10];
- bent-blade film premetering patent [11];
- patent application for a conventional short-dwell/grooved-rod premetering patent application [12];
- patented new type of coating head intended for film press and blade coating [13].

From the beginning, it was clear that this new, 'revolutionary technology', as it was named, would have a bright future in surface sizing, and since then it has also gained an extensive use in pigmenting, coating and surface treating of speciality paper grades.

Based on volumetric roll premetering, BTG developed the technology by using a wire-wound roll as the premetering element. This technology has also had good success [14].

Authors describing these new film press technologies are, for example, Frei [14], Rantanen and Finch [15], Knop and Sommer [16] and Küstermann and Hoericke [17].

2.4 Coat-weight development and metering-element-related phenomena

In the film press, metering on the film press roll and transfer to the sheet can be handled as separate single operations. Film metering is a 'simple' process in comparison with blade coating where, because of dewatering, an increase in viscosity of the coating colour as well as the surface roughness to be coated have to be taken into account.

The following four different metering processes are available for different applications [15]:

- small grooved rod;
- large-diameter smooth rod;
- large-diameter grooved rod;
- blade.

2.4.1 Grooved-rod metering

Based on the volumetric metering principle, the grooved rod is very easy to use. The coating medium is simply applied through the open area between roll and rod. The speed or viscosity has no major influence on the metered wet film thickness. Because the grooved-rod surface partly penetrates the

(a)

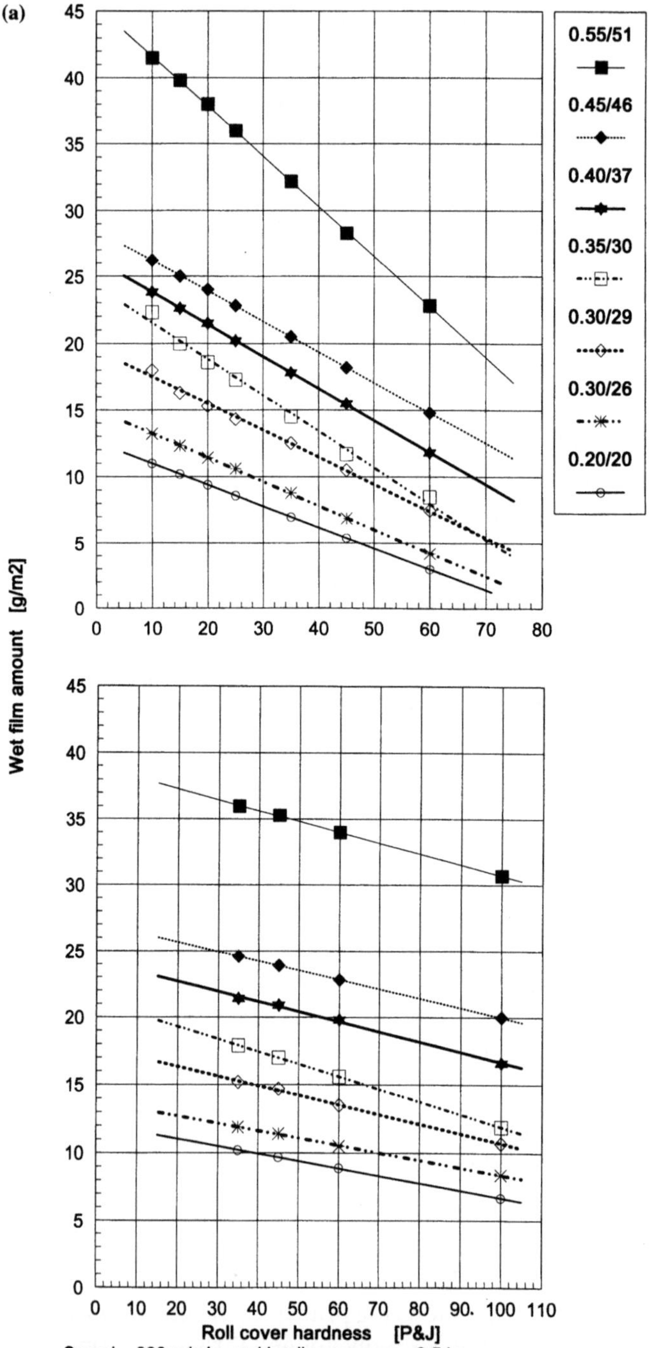

Speed 800 m/min, rod loading pressure 0.5 bar
Medium viscous oxidized potato starch

(b) Solids content 9.9 %, Br 100 rpm-visc. 29 mPas

applicator roll surface, roll hardness has quite a big influence on this open area; as also does rod loading, especially with the softest covers. Because the same P&J hardness rubber and polyurethane covers behave differently under dynamic conditions, curves of film amount against rod loading have to be made separately for these materials (Figure 2.1).

Grooved-rod wear can be a real problem. In surface sizing, excessive wear occurs when the base sheet contains hard fillers such as TiO_2, calcined clay or just filler clay. The explanation is that some of the filler pigments are transferred in the nip from the sheet surface onto the applicator roll surface and in this way come into contact with the rod. The problem can be minimized by using softer applicator roll surfaces and by running new rods with a reasonably low rod loading. When the rod wears, rod loading has to be lowered to maintain a stable size amount. During the final stage, the wear may continue very rapidly owing to fully worn chrome plating. The rods have to be changed when the film cross-profiles become unstable, causing an unacceptable moisture content profile at the reel. In pigmenting and coating, pigment type has to be carefully chosen. Fine calcium carbonates are the best with respect to wear but other pigments, e.g., delaminated clays, are also used successfully in low-solids pigmenting. With increasingly higher solids contents, the rate of wear increases rapidly. So, in high-solids coating, rod lifetime seldom exceeds four days even with the most optimized coating colours. Also, unsuitable applicator roll covers can be a reason for excessive rod wear. Covers containing hard fillers cause very rapid wear. The best covers in this respect are made of polyurethane without any fillers. In rod development, a number of new, more wear-resistant rod covers have been tested but none has reached market acceptance.

Normally, there are no major defect-related problems in surface sizing. It is possible for roll covers to become size-repellent, causing a phenomenon such as small-scale skipping. The reason can be that there are fatty unstable additives, such as defoaming agents in the size or water-repellent agents such as alkyl ketene dimer type emulsions having too low a glass transition point which causes them to soften or melt at size temperature. When running very demanding paper grades it is recommendable to have a high size film thickness and a relatively high size solids content. Especially with low-quality pearl starch, the rod bed surface sometimes becomes dirty and in extreme cases can cause streaks on the films. The reason is that this type of starch, when cooling, retrogrades, causing streaky films and misting after the rod. As a result of this risk, most mills running a film press in North America use ethylated starch, which is not that temperature-sensitive. The problem can also be minimized by having a lower starch

Figure 2.1 SymSizer surface sizing with grooved rod, with (a) polyurethane roll covers; and (b) rubber roll covers.

viscosity, higher starch temperature, warm size circulation, applicator and support beams before sizing start-up, fine enough screens and a higher rod rotation speed.

2.4.2 *Smooth-rod metering*

In smooth-rod metering, the forces acting on the rod are of the same nature as in the bent-blade type coating process. In bent-blade coating, the most important force comes from hydrodynamic pressure in the converging metering nip [18]. Another important force is the coating medium impulse force [19] acting against the metering element.

Relevant knowledge about large-diameter rod-metering can also be adapted from the paper by Sommer [20], where comparisons are made between a rigid blade, a bent blade, a small-diameter smooth rod and Hydro-Bar metering elements in direct paper coating.

The main features and related phenomena in large-diameter smooth-rod metering are as follows:

- the rod is pressed into (that is, deforms) the backing roll to a greater or lesser extent;
- the degree of deformation depends on two important factors: roll cover hardness and web speed;
- increasing nip load increases the contact zone length;
- the rod acts on a large contact zone giving a flatter coat weight curve in relation to web speed than a metering unit which acts on a small contact zone (e.g., rigid blade);
- there is no streaking tendency.

In large-diameter smooth-rod coating colour premetering, a 'film-splitting' problem may exist in the form of fine scratches or misting at the exit of the rod nip. These problems have been overcome by:

- softening the roll cover hardness;
- increasing the rod surface speed;
- reducing the coating colour solids content;
- improving the coating colour rheological properties by supplementary pigments.

The elimination of the problem by increasing the number of rod rotations per minute must indicate that the increased shear rate generated by rod surface speed builds up a boundary layer penetrating the nip exit diverging film. So, it is quite clear that film splitting must occur inside the film leaving the rod nip to eliminate scratching, misting and spraying type problems. Or, in other words, the film is not allowed to detach itself from the dry solid rod surface. Or one could say that the environment is favourable for film splitting when only the surface energies of air and coating colour are acting without the solid rod surface having an effect. This idea has also been proven by running the rod in film speed direction by which means 'rheological scratches' disappeared.

The reasons why a softer nip or a longer nip under the rod minimizes the problem have not been analyzed. Possible explanations are, among others:

- lower pressure in the diverging nip;
- non-existence of shear blocking;
- more time for orientation of pigment particles and straightening of polymers under shear forces;
- sufficient time for formation of a low viscous boundary layer on the rod surface.

Even though the large-diameter rod metering process is very different from the rigid-blade coating process, it looks like the 'bleeding' type problems have partly the same origin. Thus, for blade coating, Gane, McGenity and Watters [21] have made the statement that the effect of film splitting in the context of roll coating is well known, and the mechanisms of filamentation and cavitation suggested here are not dissimilar. Also, literature review and experimental results [22] concerning the influence of coating colour solids content and type of pigments on bleeding in blade coating seem to comply with smooth-rod premetering experiences. As in blade coating, the worst pigments to run are coarse plate narrow particle size distribution pigments.

Because there is no need for water retention in premetering, the coating colours containing only pigment and latex often have good runnability. This fact has also been shown for short-dwell coating by McGenity *et al.* [23]. To widen the operational window, given by the rod load versus coat weight relationship, viscosity modifiers often have to be used in rod premetering.

Besides coating colour properties, factors affecting the coat weight are rod loading, roll cover hardness and speed. The commonly used P&J hardness measurement does not give accurate enough information to predict coat weight development when covers are made of different materials, as shown in Figure 2.2. Use of softer covers under running conditions, the coat weight increases; but in laboratory measurements, softer rubber covers give lower coat weights than do harder polyurethane covers.

Because the first simulations using computational fluid dynamics and data from different shear rate viscometers have resulted in good correlations between calculated film amount and coat weight from pilot machine trials, the mathematical modelling will act as a useful tool for further process development work.

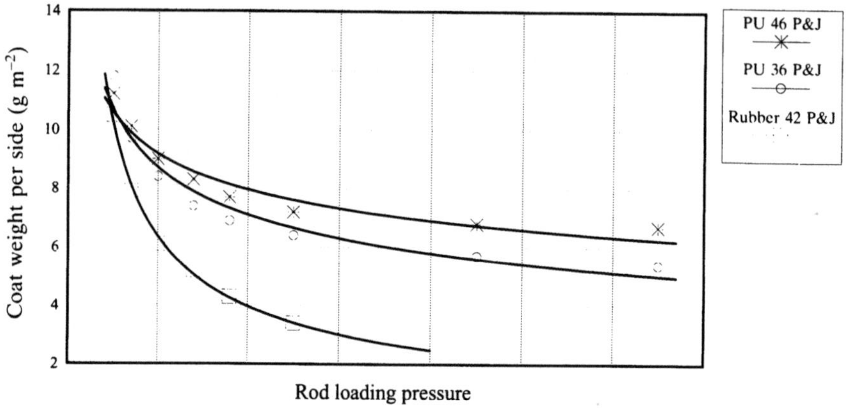

Figure 2.2 SymSizer: smooth-rod 35 mm characteristics – coat weight and rod loading pressure. Coating colour solids content ≃ 60%. Formulation (in parts per hundred): pigment:starch:latex = 100:5:8. Speed = 1100 m min^{-1}.

2.4.3 *Large-diameter grooved-rod metering*

The improvement targets in developing the large-diameter grooved rod were:

- reduce grooved-rod wear;
- produce better unstreaky films with rheologically difficult coating mediums;
- reduce roll cover scratching.

As in grooved-rod metering, speed cannot have an effect on the film thickness. But when hydrodynamic forces are high in high-solids pigment coating, the rod partly behaves like a smooth rod, so the coat weight difference between smooth-rod and grooved-rod metering is small. This means that direct contact forces between rod and roll cover are lower than in small-diameter rod metering where hydrodynamic forces are much lower. The large contact area between rod and roll cover also reduces the maximum specific pressure between rod and roll, which also reduces rod wear.

Because of longer nip and higher hydrodynamic pressure under the rod with some high-viscosity polymer mediums, better films on the roll surface have been achieved. Also, a lower nip angle between rod and roll allows the impurities to go easier through the nip, preventing streaks on the covers or clogging of the grooves.

Because of limited experience in production, it is too early to say how well the targets can be fulfilled. Until now, experiences have been positive.

2.4.4 *Blade metering*

In the traditional bent-blade-mode premetering, film profiles and scratches are problems. At zero-angle mode, if the blade tip bends excessively, it causes lining in the film (soft shallow streaks) and totally destroys the paper quality. With the double-loaded blade process, where the first tube is used for blade loading and film amount adjustment and the other tube loads the tip of the blade with no effect on coat weight, the problem of lining is avoided. This process has the longest shear forcing area under the metering element and results in good profiles and runnability. Because of a large operating range with respect to film thickness, this process may see a renaissance in surface sizing and in coating with non-abrasive coating colours. For high-speed pigment coating, blade wear problems prevent its use.

2.5 Film transfer at film-press nip

2.5.1 *Surface sizing*

Even though around 30 million tons a year of different paper grades are surface sized today with film presses, no theoretical studies have been published concerning this process. The reason is that in gate-roll metering, the size film either is very thin or has such a high viscosity that size penetration is very limited and there are no ways to change the relationship between film thickness and size viscosity. With the new type of film presses, however, it has been easy to develop and control paper quality by adjusting size film thickness, solids content and viscosity.

Theoretical aspects concerning pond size press nips are, however, important to help understand the film transfer size press operation. In the following analysis are terms derived by Hoyland *et al.* [24] for pond size press. For film presses the invalid size absorption term is disregarded. Thus, we have:

$$\mathrm{Pu} = K_1 + K_3 \eta S, \qquad (2.1)$$

where Pu is wet size pick-up;
K_1 and K_3 are coefficients;
η is the viscosity of the sizing solution;
S is the web speed.

The first term on the right-hand side of equation (2.1), K_1, is called the sheet term, and the other, K_3, the hydrodynamic term. The sheet term accounts for the sizing solution that is immobilized owing to surface roughness and any effects caused by the compression and expansion of the web as it passes the nip. The sheet term depends only on base-sheet properties.

The quantity of sizing solution left on the surface of the paper on exit from the nip is reflected in the hydrodynamic term. The coefficient K_3 is given by:

$$K_3 = \alpha \frac{5}{4} \frac{R}{p} \frac{6.5}{(2.7 - \mu)}, \qquad (2.2)$$

where α is a film-splitting factor (0.5 is symmetric splitting);
R is the roll radius;
p is the nip loading;
μ is the roll surface deformation factor.

The roll surface deformation factor, μ, is given by:

$$\mu = 2\left(1 + \frac{2R\eta SE}{(1 - \nu^2)p^2}\right), \qquad (2.3)$$

where E is Young's modulus;
ν is the Poisson ratio for the roll covering material.

These equations are important when considering factors having an influence on the maximum achievable size pick-up which is limited by the minipond formation. Because there is no converging nip full of size, which would create a high hydraulic pressure as in the pond size press process, the maximum achievable size amount by the film press with the same size is much lower than that by the pond size press. So, the size solids and/or viscosity has to be higher to achieve the same bone-dry size amount as with the pond size press.

For size penetration depth considerations, equation (2.4) below, presented by Krischer [25] is important. Liquid transported under pressure through a porous material can be described by:

$$m = \frac{f}{\mu^{\text{pore}}} \frac{b}{\mu^{\text{path}}} \frac{\mathrm{d}P}{l}, \qquad (2.4)$$

where m is mass flow;
f is the total area;
b is the flow coefficient;
$\mathrm{d}P/l$ is the pressure drop;
μ^{pore} is the pore coefficient;
μ^{path} is the path coefficient.

The pore coefficient is the ratio of the total area of the porous material to the total area of the pores. The path coefficient takes into account the increase in the flow path caused by the detours the liquid has to make when passing through the layer and the flow losses caused by expansion and constrictions of the pores.

Fine papers up to 100 g m^{-2} basis weight can have better internal bond strength when surface sized by the film press instead of by the pond size press as measured by Z-direction strength tests. Film presses normally have a 50% higher solids content than do pond size presses while the applied bone-dry size amounts are about equal. So, in this process, less size liquid penetrates to the finest pores in the sheet, and the size volume mainly fills big voids. Size liquid is not lost for strengthening the fine pore volume of the sheet, where natural fibre bonds already have created good strength properties, but instead fills relatively deeply the big voids, with high-solids size having much higher bonding power than the lower-solids size used in pond size presses. With film press, more size stays anyway on the surface and so surface strength is better on film press surface sized paper than is achieved with pond size press with the same bone-dry size amount [26].

This experience is possible to explain based on equation (2.4). On pores size flow amount is directly proportional to the pore area and so to an increase in pore size diameter to the power of 2. The flow coefficient is inversely proportional to the pore size and so the flow depth is bigger on bigger pores. Also the path coefficient evidently is bigger for smaller pores. So these factors can explain why a bigger share of a premetered size film thickness is filling the big voids in a paper sheet with film presses compared with pond size press where unlimited amount of size is available to penetrate into the sheet. This means that even though size viscosity is bigger with film press than with pond size press and the available size amount is substantially lower the big voids get enough size for strengthening. A common way to increase the size penetration with the film press is to use a lower viscosity type of size than with a pond size press.

At high speeds, film splitting after the nip creates misting when too thick a film splits at the nip exit. This inconvenience can be eliminated by lowering the film thickness, by using a lower velocity size which penetrates more completely, or by using harder and/or thinner roll covers and a higher nip load. Especially at low speeds, minipond formation limits the use of these factors.

2.5.2 *Coating*

In film press coating, the nip phenomena are even more complicated than in surface sizing. The stages in film transfer are:

- base-sheet penetration of the coating colour;
- dewatering of the coating colour;
- film splitting at the nip exit.

Coating colour penetration. Too deep sheet coating colour penetration is a problem especially with film coating of high-porosity fine papers. One test to evaluate penetration depth is the burn-out test by Dobson [27] where the greyness level after burning directly gives an indication how deep the pigments are penetrated. The other test, *Z*-direction ash content analysis, explicitly shows the coating colour distribution. This measurement is accurate when the base sheet has no filler such as is often the case in trials run on newsprint base. The Cobb–Unger$_{10}$ (CU_{10}) oil absorption test has shown a good correlation of coating colour penetration depth. Base sheets having a low CU_{10} value give better quality coating with respect to smoothness, paper gloss and printed gloss.

- speed = 20 m s^{-1};
- nip length = 40 mm;
- penetration depth = 25 μm;
- pore radius = 10 μm;

then the average value throughout the nip length for flow velocity can be calculated as 10 m s^{-1} and for shear velocity as 10^3 s^{-1}. Taking into account that coating colour penetration as a bulk can maybe occur only over a few millimetres inside the nip, the real shear velocities are nearly comparable to shear rates under metering elements.

Blade-coating fibre coverage is reviewed and studied by Salminen and Fors [28]. The results show that with standard latex/CMC (carboxymethylcellulose) offset grade coating colour, penetration could not explain differences in smoothness caused by solids content. Higher coating solids in the studied system probably improved surface quality mainly by lowering base-paper expansion rather than by affecting coating penetration and coating layer bulk. Coating colour penetration was measured only through using pigment–latex-based colour without water-retention agents.

Coating colour spreading. Studies made by Gregersen, Johnsen and Helle [29] have shown how water-based flexo inks penetrate significantly into the interfibre voids of the surface layer. Ink squeezes away from fibre peaks and fairly large voids are filled with ink. It is very obvious that lateral coating colour flow has an important effect also in coating and surface sizing with a film press.

Coating colour de-watering. De-watering of coating colours in contact with paper has been one of the main subjects studied during past years, first for blade coating and now also for film-press applications. According to filtration theory (Figure 2.3), a filter cake or immobilized layer consisting of the pigments and the non-soluble components of the coating colour is formed at the interface between paper and coating colour [30]. Also in this case, liquid transport through a porous material, filter cake,

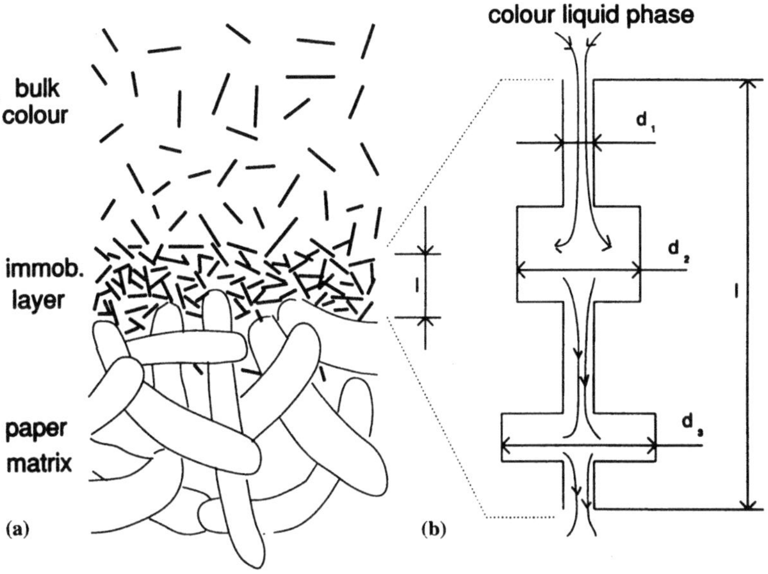

Figure 2.3 Schematic view of the situation at the colour–paper interface during filtration de-watering.

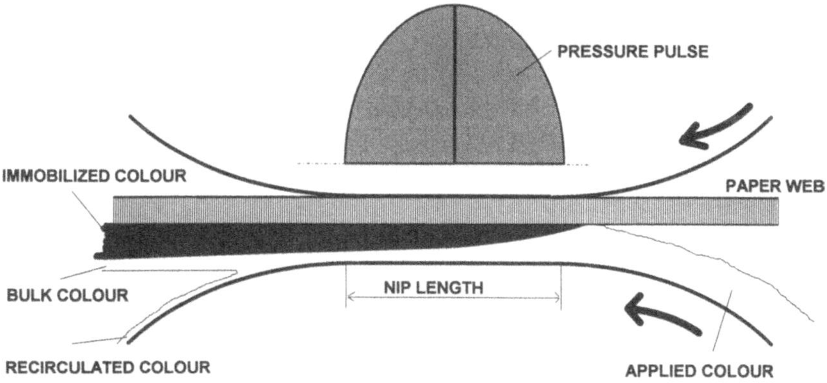

Figure 2.4 The de-watering in the film-press nip [30].

can be described by equation (2.4) [30]. The de-watering of a coating colour in the film-press nip is shown schematically in Figure 2.4 [30].

If the filter cake de-watering mechanism is assumed to be valid, the de-watering of a coating colour can be described mathematically according to equation (2.5) [30]:

$$\frac{m}{f} = (CA\Delta Pt)^{1/2} , \qquad (2.5)$$

where m is the de-watered liquid phase amount;
f is the de-watering area;
A is the colour parameter;
C is the structure parameter;
t is time;
ΔP is the prevailing pressure.

The colour parameter A includes information about the application and immobilization solids contents of the colour, and the structure parameter C gives information about the viscosity of the liquid phase and about the packing structure of the immobilized layer. For simple systems, the parameters A and C can easily be calculated; for industrial colour systems, these parameters can be determined experimentally.

Based on these theories, Letzelter and Eklund [30] have calculated the effects of the main parameters on the de-watered liquid phase amount and the thickness of the immobilized layer. Large filter cake layer thickness is achieved by having a low viscosity liquid phase, big particle size pigments, low packing volume at a given porosity, high-solids content of the coating colour, high pressure in the film-press nip and low machine speed. The applicator roll radius or roll cover hardness does not have any effect on the de-watering in this model.

Another new presentation based on similar modelling and his own experimental work is presented by Trefz [31].

The first experience with really soft applicator roll covers came from the BTG TWIN-HSM process [32]. With softer covers, there is less coating colour penetration and the coating colour is less de-watered.

Film splitting at nip exit. It is assumed that at the exit of a film-press nip, the film splits in the middle of the bulk coating layer left between the immobilized layer and the applicator roll. Film splitting has been extensively studied in printing technology, which is quite a similar process in this respect. De Grâce, Dalphond and Mangin [33] have considered film splitting as a four-step sequence:

1. cavitation;
2. filament formation;
3. filament elongation;
4. filament rupture.

Even though formation of the immobilized layer has not the same importance as in film-press technology, the film splitting itself follows the same rules.

Brockmeier and Emilsson [32] classify film split on a film press in two categories:

- one caused through excess coating applied on the surface;
- one caused through dilatancy.

Means to avoid film split are:

- press premetered coating amount into the sheet body;
- apply less;
- avoid dilatancy at the transfer point by a low specific pressure in the nip.

Letzelter and Eklund [30] have made experimental work with a device called a web–peel meter (Figure 2.5) to study the interactions between coating colour, paper and cover material in relation to film-splitting phenomena. Their conclusions were that the adhesion or cohesion force does increase, even when the whole coating layer has reached the immobilization point. Thus, the de-watering of the coating colour above the immobilization point is not beneficial and the measured adhesive forces are primarily determined by the cohesive forces in the colour. The solids content of the coating colour itself is probably not the decisive parameter, instead it must be properties that are dependent on the solids content, as, for instance, viscosity. It is also conceivable that viscoelastic properties, such as the elongation or extensional viscosity, could be of importance for the development of the film-splitting location.

The investigations presented here also suggest that the most preferable situation at the nip exit must be one where a thin film of coating colour with application solids content is present at the interface between coating colour and roll surface. Because the solids content and the cohesion force in this film are low, film splitting will thus take place in this film. As a consequence, only a thin film will be recirculated by the film press roll.

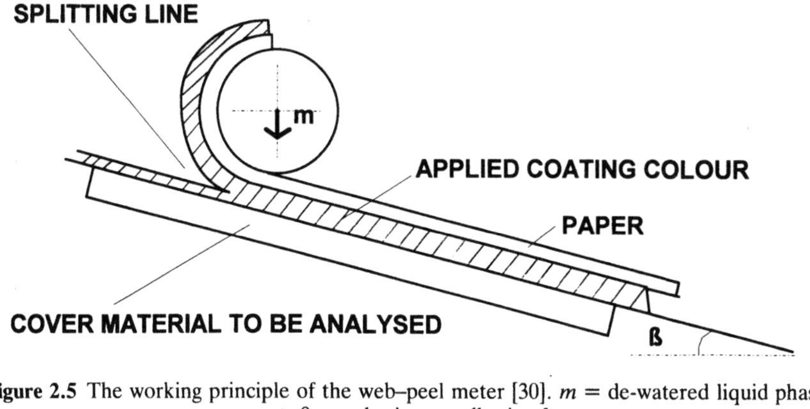

Figure 2.5 The working principle of the web–peel meter [30]. m = de-watered liquid phase amount; ß = cohesive or adhesive force.

It is therefore obvious that the de-watering properties of the colour, the pressure conditions in the nip as well as the absorption ability of the paper are important parameters in determining the location of the film split.

Web stealing. After the film-press nip, the web should peel steadily from the roll, but in web stealing the web transfers from one roll to the other very rapidly thus causing patterns in the coating. These patterns can be seen in the coated paper with help of ultraviolet light and optical brighteners. The film splitting on the roll, which the web follows, takes place along a line such as at the exit of the paper machine press section roll and, on the opposite side, the film splitting direction is perpendicular to the web surface. These geometrical differences in the film splitting produce differences in the film transfer factor and differences in the coat weight amount. As with all size presses, the web has the tendency to follow the harder roll with higher speed and the side where the coating medium has a higher viscosity.

Theoretical considerations based on offset printing technology. 'Tack' is a term used widely to describe forces or energies involved in the separation of two surfaces joined by a thin film of liquid. It can be defined as the maximum tensile stress that the liquid medium can withstand before rupture. Tack is a viscoelastic property. The length of the filaments at rupture may also be considered to be a measure of the elasticity or shortness of an ink [33].

Besides being a function of ink composition, tack is also affected by ink film thickness. Stefan's law expresses the force needed to separate two plates immersed in a fluid as a function of the velocity of separation, the fluid viscosity, the area in contact and the cube of the fluid thickness between the two plates [34]:

$$F = \frac{v^{s} \eta f^{c}}{h_x^3}, \tag{2.6}$$

where F is force;
v^s is velocity of separation;
η is fluid viscosity;
f^c is contact area;
h_x is fluid thickness.

In film-press coating, the web has a tendency to follow the side which is smoother, has less absorbency or has a higher film thickness. This means that the free splitting film is incomplete because, based on Stefan's law, a thicker film should split with lower force.

In cases where a thick film is immobilized onto one of the roll surfaces, the web follows this side.

2.6 Precalendering

Especially for all coated paper grades, precalendering has to be studied case by case. Precalendering always improves the paper smoothness and gloss of coated sheet with low coat weight. To avoid orange peel mottle, there is an optimum precalendering rate. The decision to have none or one-nip hard–hard, hard–soft, or soft–soft precalendering, or two-nip soft precalendering depends on the quality targets, uncalendered base-sheet properties, coat weights and the process of final calendering. In most cases, investment costs, required space and also efficiency factors play an important role in the final decision-making.

2.7 Control of paper quality

To achieve the best runnability and paper quality, the film-press conditions have to be optimized for different applications.

In coating as a rule, as low a nip load as possible has to be used to achieve the best fibre coverage. Even though, theoretically, roll hardness has no effect on coating colour pressure de-watering, at very low speeds as soft as possible up to 100 P&J covers have to be used in combination with as low as possible linear load to eliminate coating colour drying on the roll surfaces when using low-water-retention coating colours.

On the other hand, at high speeds with coating colours having a high water retention, the problem is that not enough de-watering takes place and so a splitting of thick bulk coating colour layer originates heavy misting and bad orange peel phenomenon after the nip. Based on pilot experiences, harder roll covers about 30 P&J (polyurethane) are needed to achieve enough de-watering. So, it is quite evident that de-watering in the film-press nip is normally more dependent on maximum surface pressure than on time at all speeds.

Practical experience has shown that coated paper has the best fibre coverage, smoothness and a faultless surface when the maximum solids content for each coating colour is used. Exceptions are those cases where, in the nip, the immobilized layer has extended throughout the thick film to the applicator roll surface and then split; or where defects are coming from rod misting. With low coat weights, rheologically easy coating colours or very optimized conditions the film transfer factor can be nearly 100% without causing problems related to drying on the roll surfaces. This means that immobilization right to the roll surface is not always dangerous but can even be beneficial.

Nip behaviour can be explained by the theory that a low thickness splitting bulk layer of coating colour is unable to produce large enough filaments capable of generating quality or runnability problems. So, de-watering control has a crucial importance both with respect to orange peel and nip misting phenomena.

Besides film-splitting thickness, filament formation and the influence on paper surface quality depend on many other factors such as base-paper smoothness and porosity evenness, filament strength, filament breaking point, fluidity and spreading of detached filament attracted back to the web, surface tension of the colour, etc.

For paper quality and coating colour runnability quite the same rules apply as in blade coating. The biggest difference possibly concerns runnability. Coating colours impossible to run with blade coaters at high speeds can give trouble-free operation in smooth-rod metering and excellent runnability with respect to misting at the exit of a film-press nip.

Pigment choice is mainly determined by the coated sheet quality targets and by cost. The effect of coating colour pigments and binders on the runnability and the paper properties is basically quite similar as in other coating processes. Pigment choice is essential in adjusting the coating structure and runnability properties. So, operation window adjustment can be performed by changing coating colour rheological and water retention properties by optimization of pigment particle size, particle size distribution, form factor or pigment mix. If only pigments and latex can be used as components in the coating colour, the result in both coated sheet quality and high speed (1200–2000 m min^{-1}) total runnability is often very good.

At low speeds (300–600 m min^{-1}) high shear synthetic viscosity modifiers often have to be used in smooth-rod premetering to reach sufficient coat weight adjustability. To eliminate coating colour drying on the applicator roll surfaces, low viscosity type CMC is normally used. A small amount of CMC addition may have a positive effect on the paper quality by preventing excessive coating colour penetration inside the sheet.

Because all additives in coating colours increasing liquid phase viscosity and/or interparticle bonds may have an adverse influence either on runnability or coated sheet quality, the amount of all additions has to be studied carefully on a pilot scale before taking into production.

Starch is very much used as a binder in film-press coating, but sometimes it cannot be used at the desired high amounts because of runnability problems at very high speeds and/or when coating quality targets are very high. So, there is a demanding challenge for starch manufacturers to develop new starch types better suited for future film-press coatings.

2.8 Nip mechanics and roll covers

Size press requires a straight nip load. In the high-speed pond size process, deflection-compensated rolls were used to fulfil this requirement. On these size presses at high speeds, high hydrodynamic forces needed nip loads of the level 50–60 kN m^{-1} to eliminate excessive pick-up or slipping on the nip.

On film presses, operating nip loads occupy a wide range of values (10–40 kN m^{-1}) depending mainly on the runnability and the effect on the paper quality. Nip load is no longer determined by danger of excessive pick-up.

The demands for even nip load have not decreased in any way. Nip load has a direct influence on film transfer factor. A higher load increases pick-up, which is the opposite of the normal size press operation at high speeds. So, if the sheet edges are dry on the reel, increasing the nip load or decreasing the crown corrects this problem when coming from uneven nip load. With respect to wrinkling or web tension profile, the rules are the same as in pond size press operation. So, with too low a crown, the sheet is loose in the centre and tight at the edges. In this case, linear load lowering would correct the problem.

Especially with low linear loads under 25 kN m^{-1} often used with covers softer than 25 P&J, slipping in the nip may cause tension and wrinkling related problems. One way to eliminate these problems is to run with both rolls having speed-controlled drive instead of master roll and torque-driven roll. From this way of running, experience today is very limited, but good. Because the hardness difference between the applicator rolls also causes slip between the rolls inside the nip, to achieve a static friction situation, covers having the same hardness are commonly used. This requires a straight take-off angle from the nip exit so that the sheet can follow the roll, which it has the bigger tendency to follow. With harder, about 20 P&J, hardness roll covers and higher linear loads, about 30 kN m^{-1}, slipping and crown-related problems very seldom arise.

An uneven moisture content profile comes from size press operation or other factors. To check for proper film press operation, film amounts are measured on both edges and in the middle of the rolls on both sides before the nip and at least on one side on the exit side. If all profiles are straight, the problem comes from some other source than the film press. This measurement is done by film scraping and the gravimetric weight method and also gives in both coating and surface sizing the film transfer factor, which is very important to know with respect to both runnability and sheet quality.

Roll cover choice is one of the most important issues when investing in a film press. Grooved-rod metering requires non-abrasive and not too hard covers, and smooth-rod metering has strict demands for a maximum allowable hardness to guarantee good profiles, but the cover is allowed to contain hard abrasive fillers to minimize roll cover wear and scratching. Today polyurethane and rubber covers are both used with all metering elements.

2.9 Quality differences between film-press and blade coated sheets

Film-press coating has an advantage in quality when brightness has to be increased by coating or when very low coat weights of under 6 g m^{-2} per side have to be run. Contour-type coating also has a low tendency to mottle print. Blade coaters have the benefit of creating better smoothness coupled with the ability to run high coat weights. With low basis weight wood-containing grades, paper gloss and other paper or print quality characteristics are at the same level when the comparison is made with the same coat weight.

Because of the better runnability, the film press can use base sheets that are weaker or may contain more impurities than is the case with blade coating. Either a higher filler content up to 25% or a lower amount of kraft pulp in the furnish of wood-containing grades is used without sacrificing the efficiency in papermaking or in printing. Both measures improve the film-press related weak point, smoothness. Film-press-coated wood-containing sheet has much less tendency to fibre roughening in heat-set offset printing than has a blade-coated sheet. This fact compensates even more for the poorer original smoothness. The explanation of this phenomenon is not clearly understood, but higher coating layer porosity and contour-type coating, which better binds the fibres on the surface, may be the explanation. A film-press-coated sheet has good printing runnability with respect both to web breaks and to blanket washing intervals. Use of a film press makes the base-sheet defects, such as holes, less detrimental. Blade coating can make the defects worse.

2.10 The SymSizer film press

2.10.1 Process designs

Important steps in the development work have been:

- 1987: a new coating head pressure chamber design (Figure 2.6) [33];
- 1990: large-diameter smooth rod [35];
- 1991: double-loaded blade [36];
- 1992: large-diameter grooved rod [37];
- 1992: C1S + C1S layout (Figure 2.7) [38].

Some other mechanical/process SymSizer features are:

- small and big rods go into the same space;
- no skipping or coat weight instability even at high speeds (1500–2000 m min^{-1});
- the same pressure chamber principle used in Optiblade [39];
- low circulation rate, supply to the beam 80–100 l m^{-1} min^{-1};

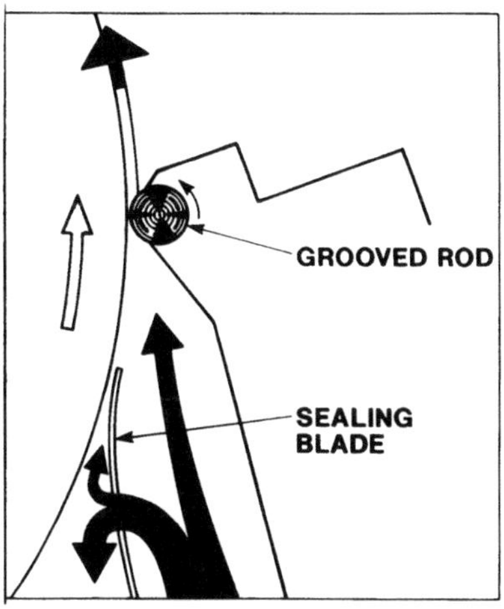

Figure 2.6 SymSizer sealing blade principle.

Figure 2.7 Sym Coat C1S + C1S layout.

- straight nip load, beam loading forces equalized on both ends;
- ability to run very low nip loads by hydraulic loading;
- the big beam for large machines rotates around its centre point instead of retracting, giving:
  - high positional accuracy;
  - space savings;
  - safe operation;
- minimum number of moving parts and wearing parts;
- high priority given to ease of cleaning and maintenance.

The newest development work has resulted in:

- new design where all functions are integrated, i.e., frames, walkways, instrument cabinets (Figure 2.8):
  - reassembly not needed;
  - transportation in one piece;
  - installation time and work on rebuilds minimized;
- web upwards design for minimized space requirement;
- new modularized applicator head design:

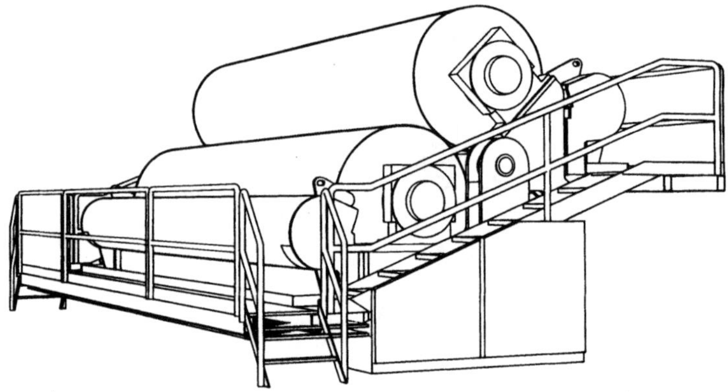

Figure 2.8 SymSizer design.

- functional part of the applicator head separated from applicator beam;
- the low temperature control in the beam results in sweating of the beam, thus leading to clean and safe operation;

- new temperature equalizing system of rolls:
  - better temperature profile of rolls since water is fed to the rolls through the front journal and removed through the drillings of back side head by centrifugal force;
  - no possibility for water load;
  - no pressurized air needed for removing condensed water;
  - easier maintenance as there are no wearing parts in front side water joint.

2.10.2 Applications and production experiences

With its flexibility not to compare with any other coating process the film press has gained a wide acceptance in the production of almost all paper and board grades. For example, of the 67 SymSizers in production (February 1996), 17 are coating with high solids content, many are changing to coated grades and among the other sold references more than 10 are for coating purposes. Hot topics today are high-speed coating, high coat weight coating of wood-free and board grades, new generation office papers, speciality fine papers and different recyclable barrier coatings. All these are among the newest film press applications.

Depending on the application concerned, the following surface sizing or pigmenting methods are used with SymSizer:

- pond type surface sizing with grooved rods;
- gate-roll type high solids content low film thickness surface sizing either with grooved rods or large-diameter rods;

- gate-roll type low solids content low film thickness surface sizing either with grooved rods or large-diameter rods (excellent runnability in surface sizing of newsprint);
- low solids content pigmenting either with grooved rods or large-diameter rods;
- gate-roll type high solids content pigmenting with large-diameter rods;
- clear water application for decurling.

For coating, large-diameter smooth or grooved rods are today used in the speed range of 250–1250 m min^{-1}, coat weight range 1.5–15 g m^{-2} per side and solids content 45–65%.

In surface sizing, the efficiency and energy requirement targets have been achieved without sacrificing paper quality. As a rule, film-press surface sized fine paper is even better than pond mode surface sized paper with respect to smoothness, cockling tendency, absorption evenness, profiles, etc. The bottleneck in speed has been eliminated and the number of breaks has decreased by 60–90% in comparison with pond surface sizing. Because of higher incoming moisture from the pre-dryer section, a low number of breaks and high size solids content, the total energy requirement dropped 15% per produced paper ton on the first reference where these calculations were made [40].

For film-press coating applications, the interest has grown explosively during the past couple of years. The first SymSizer coating reference started coating at the end of 1989. The metering device there for coating has been bent-blade followed by double-loaded blade and then large-diameter smooth rod.

MoDo Husum papermill PM 6 started fine paper precoating production in early 1992 and was the first to use large-diameter smooth-rod premetering.

Machines producing wood-containing coated printing grades have efficiencies up to the world record class and the paper quality is very competitive. One machine runs with good results 100% de-inked pulp furnish for surface sized and coated fine paper printing and office grades. In light weight coat (LWC) production, both one-station SymSizers and two-station Sym Coats are working without problems on wide high-speed machines, so it is too early to make an exact comparison between these processes. Both have their own significant benefits.

In coating, the average roll cover regrinding interval is over six weeks, smooth-rod lifetime over four weeks and rod-bed lifetime over four weeks. On most machines, the average number of breaks in coating is about one break per day in the SymSizer area.

As in surface sizing, base-sheet moisture content has no effect on the coating profile, so the moisture content before the film press can be high, normally 4–7% in coating. An important economic factor is also that very little expensive contactless drying is needed before the cylinders in the after-dryer section. One machine runs 850 m min^{-1} with 10 g m^{-2} per side coat weight on both sides having only three infra-red units. Normally, 80% dry content on the sheet surface is enough to eliminate cylinder and fabric staining. It is also possible to have very high-intensity drying in a short space without danger of developing the uneven absorption properties responsible for print mottle. This comes partly from contour-type even coat weight profile and coating colour de-watering in the nip. For runnability reasons and to reduce the risk of excessive sheet surface temperature, the air foil dryer has increased its popularity as the basic contactless dryer in film-press coating applications. Previous problems related to runnability with air-turning devices have been eliminated by intensive development work by Valmet. This development has also resulted in working web tension control by using air-turn control parameters.

References

1. Booth, G.L. (1993) *Transfer Roll Coaters, A History in the Coating Processes* (ed. J.C. Walter), Tappi, 115–23.
2. Hansson, J.A. and Klass, C.P. (1984) High-speed surface sizing. *Tappi J.*, **67** (1), 64–6.
3. Westergård, S. and Rantanen, R. (1983) Twostream – eine neue vielscitige Streichanlage. PTS-Streicherei-Symposium, 102–7.
4. Pratt, G.W. and Muenchow, H.L. (1990) Pilot studies using a hydrophilic roll coater to surface treat groundwood paper with specialty starches. Tappi 1990 Coating Conference, 173–7.
5. Tehomaa, J., Palokangas, E., Mäkimattila, J. and Tuomisto, M. (1992) A comparison of techniques for surface sizing of fine papers. *Tappi J.*, **75** (8), 79–84.
6. Hoppe, J. and Baumeister, W. (1987) Betriebserfahrungen mit der Leimpresse an einer Hochleistungs-Feinpapiermaschine. PTS-Streicherei-Symposium, 124–8.
7. Bailey, D.F. and Bown, R. (1990) The use of pigments at the size press – a European view. Tappi 1990 Papermakers Conference, 199–208.
8. Rantanen, R. (1991) Pigmentieren mit der Filmleimpresse. PTS-Streicherei-Symposium, 119–23.
9. Rantanen, R. (1980) FI 800029, FR 2, 507, 108.
10. Eklund, D. (1980) FI 66448.
11. Alheid, R.J., Carlson, R.W. and Garde, G.R. (1981) GB 2103115A.
12. Woehrle, A. and Sollinger, H.-P. (1984) US 4, 848, 268.
13. Rantanen, R. and Lummila, M. (1987) FI 875333, EPO 319503.
14. Frei, H.P. (1993) Gate-roll – und Blade-Metering-Filmpressen – zwei unterschiedliche Systeme mit gleicher Zielsetzung. *Wochenblatt für Papierfabrikation*, **10**, 390–3.
15. Rantanen, R. and Finch, K.W. (1994) Production proven capabilities in SymSizer pigmenting. *Pulp & Paper Canada*, **95** (9), T 370–3.
16. Knop, R. and Sommer, I.L. (1995) Produktionserfahrungen und Versuchsergebnisse beim Einsatz der FILMPRESS als Streichaggregat. *Wochenblatt für Papierfabrikation*, **10**, 436–43.
17. Küstermann, M. and Hoericke, H.-W. (1995) Erfahrungen aus dem Einsatz neuer Streichkonzepte bei der LWC-Herstellung. *Wochenblatt für Papierfabrikation*, **22**, 1033–6.
18. Gartaganis, P.A., Cleland, A.J. and Waireki, T. (1977) Blade mechanics of extended blade coaters. Tappi 1977 Coating Conference, 145–55.
19. Eklund, D. and Kahila, S. (1978) Die Vorgaenge unter dem Schaber beim Glaettschaber-streichen. *Wochenblatt für Papierfabrikation*, **17**, 661–5.

20. Sommer, H. (1990) Influence of metering element geometry on coat quality. Tappi 1990 Blade Coating Seminar, 57–62.
21. Gane, P.A.C., McGenity, P.M. and Watters, P. (1992); Factors influencing the runnability of coating colours at high speed. Tappi 1992 Coating Conference, 117–32.
22. Suontausta, O. (1993) Stalagmite formation in LWC-Paper coating. Tappi 1993 Coating Conference, 97–105.
23. McGenity, P.M., Gane, P.A.C., Husband, J.C. and Engley, M.S. (1992) Effect of interactions between coating colour components on rheology, water retention and runnability. Tappi 1992 Coating Conference, 133–8.
24. Hoyland, R. Howarth, P., Whitaker, C. and Pycraft, C. (1977) Mechanism of the size press treatment of paper. *Paper Technology and Industry*, **18** (8), 246–50.
25. Krischer, O. (1956) *Die Wissenschaftlichen Gundlagen der Trocknungstechnik*, Springer, Berlin, p. 160.
26. Bergh, N.-O., Glittenberg, D. and Weinbach, H. (1988) Oberflächenbehandlung von Papier mit Leimpresse und Filmpresse. *Das Papier*, **42** (10), V40–7.
27. Dobson, L.R. (1975) Burn out, a coat weight determination test. Tappi 1975 Coating Conference, 123–131.
28. Salminen, P. and Fors, S. (1992) Fundamental approaches for optimizing fibre coverage in blade coating. Tappi 1992 Coating Conference, 7–36.
29. Gregersen, Q.W., Johnsen, P.O. and Helle, T. (1995) Small scale topographical variations of newsprint surfaces and their effects on printing ink transfer distribution. *Journal of Pulp and Paper Science*, **21** (10), J 331–6.
30. Letzelter, P. and Eklund, D. (1995) De-watering of coating colours in the film press. PTS Streicherei-Symposium, München.
31. Trefz, M. (1996) Theoretical aspects and practical experiences for film coated offset grades. *Tappi J.*, **79** (7), 223–30.
32. Brockmeier, G. and Emilsson, P. (1993) Betriebserfahrungen mit Twin-HSM bei der Oberflächenveredlung von Rohpapieren auf Altpapierbasis. *Wochenblatt für Papierfabrikation*, **21**, 886–90.
33. De Grâce, J.H., Dalphond, J.E. and Mangin, P.J.A. (1988, 1991) A mechanistic approach to ink transfer, Parts II, III, in *Advances in Printing Science and Technology* (ed. W.H. Banks), Pentech Press, London, vol. 19, pp. 146–61; vol. 21, pp. 312–27.
34. Mangin, P.J. (1991) A critical review of the effect of printing parameters on the linting propensity of paper. *Journal of Pulp and Paper Science*, **17** (5), 156–63.
35. Rantanen, R. (1990) FI 91299, EPO 0454643.
36. Rantanen, R., Lummila, M. and Korhonen, H. (1991) FI 912260, US 5, 286, 526.
37. Rantanen, R., Korhonen, H., Lummila, M. and Lintula, T. (1992) FI 93885 EPO 93850110.
38. Rantanen, R. and Vikman, K. (1992) FI 94883, EPO 931172548.
39. Korpela, M.S., Malkki, R., Linnonmaa, J. and Koskinen, J. (1995) Optiblade – Neuer Streichkopf für LWC bei hohen Geschwindigkeiten. *Wochenblatt für Papierfabrikation*, **22**, 1027–32.
40. Norberg, L.-O. (1990) Operating experiences with the SymSizer at MoDo's Husum Paper Mill. *Valmet Paper Machine Days*.

3 Synthetic latex binders for paper manufacture

G. BACQUET and J.-C. ISOARD

3.1 Introduction

Latex (which in Latin means 'fluid') is a viscous, milky sap secreted by certain plants such as the Brazilian hevea, or rubber tree (*Hevea brasiliensis*). By flocculating latices of various vegetal origins, it was first applied industrially to obtain rubber, used in the beginning to waterproof fabrics and to make elastic cloths. The discovery of the vulcanization process in 1846 made it possible to mould a multitude of objects. Not until 1880, when it was applied to bicycle and automobile tires, did the production of natural rubber really take off. As the automotive and tire industries expanded, the supply of natural latices could no longer meet market demand. Moreover, during the Second World War the harvest of natural latex, chiefly in Indonesia and Malaysia, dropped from 1500 thousand tons to around 200 thousand tons. The industrialized countries therefore stepped up production of natural latex substitutes. These synthetics are dispersions made up of synthesized polymer particles (based on styrene and butadiene) maintained in aqueous phase suspension. On account of their physical resemblance and similar properties the noun 'latex' was used for all these synthetic dispersions.

The chemical industry became interested in synthesizing latices toward the end of the 1930s. Their researchers worked at first on vinyl-acetate-based compounds, whose production expanded at the end of the Second World War in the fields of adhesives and paints. This activity then diversified and developed in such different areas as additives for building materials, textiles, medicine, and, of course, functional or non-functional coating and surfacing of papers and paper boards.

Today the latices used in paper applications are mainly based on vinylic monomers [1, 2], such as butadiene, styrene, acrylics and acetates. These synthetic latices appear as a white, fluid liquid composed of polymer particles suspended in aqueous phase. These particles are between 0.1 and 0.5 microns in size, depending on the properties sought.

Certain characteristics and properties of papers and cardboards manufactured with these latices result from the chemical nature of the polymer and also from the initial dispersed state; i.e., they depend on the stability and the viscosity of the latex and on its aptitude to form a homogeneous polymer film and to wet pigments. We shall develop these four different points with respect to the constraints inherent in on-machine paper manufacturing after describing how these latices are synthesized.

3.2 From molecule to particle: latex synthesis

The process that can be used on an industrial scale is the technique of radical emulsion polymerization. Before describing this process the basic principles of radical polymerization and the types of monomers employed must be recalled.

3.2.1 Radical emulsion polymerization

The synthesis of latices is based on the principle of radical emulsion polymerization. This polymerization is carried out in a reactor generally containing monomers, water, one or more stabilizing emulsifiers, a water-soluble radical polymerization initiator and modifiers. The initiator is a chemical compound that decomposes when heated (between 50–100°C), introducing into the aqueous phase unstable free radicals that trigger polymerization by acting on the monomer molecules. The most commonly used radical initiator is persulfate (of potassium, sodium or ammonium) $S_2O_8^-$,which decomposes to give $\cdot SO_4^-$ radical ions.

When heated, the radical initiator is transformed into a radical R·. This phase is called polymerization initiation:

$$R-R \rightarrow 2\,R\cdot$$

The radical then encounters a monomer molecule, M, and transforms it into a radical.

$$R\cdot + M \rightarrow R-M\cdot$$

This new radical will react with another monomer molecule to give a new radical that is longer by one monomer unit. Generally speaking, a radical containing n monomer units can react with a monomer to give a radical containing $n + 1$ monomer units. This phase is called polymerization propagation. For $n = 1$:

$$R-M\cdot + M \rightarrow R-M-M\cdot$$

For $n = N$:

$$R-(M)_n - M\cdot + M \rightarrow R-(M)_{n+1} - M\cdot$$

This chain reaction continues until a terminating reaction stops the chain by deactivating the radical. This terminating reaction is brought about by one radical acting on another. There are two types of terminating reactions. It may be naturally caused by the presence of terminating radicals created in the reaction medium; or it may be induced intentionally by introducing the appropriate radicals at a key moment of the process. This phase is called termination. The natural process may be represented as:

$$R-(M)_n-M\cdot + \cdot M-(M)_m-R \rightarrow R-(M)_{n+m}-R$$

$$R-(M)_n-M\cdot + \cdot R \rightarrow R-(M)_n-R$$

Intentionally induced termination by the radical R′ may be represented as:

$$R-(M)_n-M\cdot + \cdot R' \rightarrow R-(M)_n-R'$$

Three models are now accepted to describe radical emulsion polymerization. Their relevance depends on the operating conditions and the physicochemical characteristics of the species in the reactor [3]. The Hawkins micelle model, taken up by Smith and Ewart, is the oldest but still the most relevant one for syntheses that take place using a low-water-soluble monomer (such as styrene and butadiene) and enough emulsifier for micelles to be present in the reaction medium (Figure 3.1). When this type of medium is stirred, the organic monomer is dispersed in droplets a few dozen micrometers in diameter (entity (A) in Figure 3.1).

Part of the monomer also dissolves in the aqueous phase (B) and spreads out though the micellar phase. The micelles have a voracious appetite for organic compounds and thus trap the monomers. The molecules of emulsifier spread through the aqueous phase, among the micelles where they are associated with each other, and at the monomer–water interfaces.

Actually, the monomer and emulsier molecules are not found in fixed locations; they are in dynamic equilibrium and constantly in transit through the aqueous phase from one site to another. When heat causes the initiator to decompose into radicals that encounter the monomer molecules dissolved in this same aqueous phase (B), each radical ion joins to any

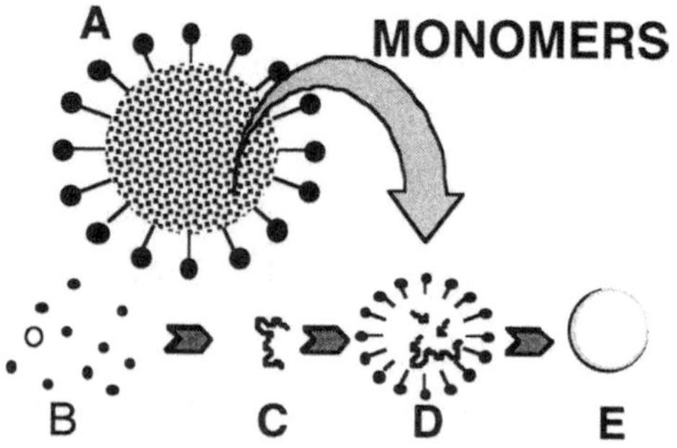

Figure 3.1 Hawkins' model. A = droplets of monomers; B = monomer molecules dissolved in the aqueous phase; C = macromolecule in the aqueous phase; D = micelles of macromolecule; and E = final latex particles.

monomer molecule that it comes across and transforms it into a radical which acts on another monomer molecule to yield a new radical that is longer by one monomer unit. This chain reaction goes on until it is stopped by a termination reaction that deactivates the radical. During growth the macromolecular ion very rapidly takes on a molecular structure that has surface active properties (C). We get the structure of an anionic emulsifying molecule with an ionized polar head deriving from the initiator and a hydrophobic tail made up of the monomer units. As the micelles are much smaller than the droplets, the growing radicals formed in the aqueous phase almost immediately penetrate the micelles, where they continue to grow through addition of the monomer present at these sites (D).

The nucleated micelles capture a radical, thus becoming polymerization sites. The monomer consumed is constantly replaced by monomer from the droplets in transit through the aqueous phase A to D. When a second radical penetrates a micelle already containing a growing radical, the reaction stops. As the micelles are transformed into particles and grow by consuming monomers, additional emulsifer (from micelles that have not yet been nucleated and gradually disintegrate) saturates their surface, thus ensuring their protection. In this kind of process the micelles are either polymerization sites or stores of emulsifier, and the monomer droplets are merely reserve pools of monomer (A).

3.2.2 The monomers

According to the description of the principle of radical polymerization, the monomers used must react with a radical molecule to give another radical molecule. This is possible if there is a double bond which will open to create a molecular bond on one side and a radical on the other. The chief monomers employed to prepare coating latices are butadiene, styrene, acetate, acrylates and acrylonitrile.

In order to obtain a range of polymer films of variable rigidities it is necessary to react together monomers of different rigidities. Generally, a monomer such as styrene or vinyl acetate, which yields hard polymers, is associated with a monomer such as butadiene or butyl acrylate, which gives soft polymers. The proportions of each monomer used are adjusted to produce the required rigidity. In this way, three large classes of latices are formulated:

- vinyl acetate–acrylate (methyl, butyl, etc.);
- styrene–acrylate (methyl, butyl, etc.)
- styrene–butadiene.

Terpolymers, consisting of three of the above-mentioned monomers, can also be made.

3.2.3 *Finishing and final characteristics*

At the exit of the reactor the latex is treated to eliminate any incompletely reacted monomers and by-products of the polymerization reaction. Purified of residual organic volatiles, the latex is filtered to eliminate any polymer coagulum that might lower the latex's user value. Finally, the solids content and the pH are adjusted according to the required specifications.

At the end of these finishing steps the latex appears as a stable milky white liquid with the following characteristics:

- solids content between 45–50% by weight;
- pH between 4–8;
- particle size between 0.1–0.2 microns;
- glass transition temperature between −30–+30°C (Section 3.6.2);
- the surface tension is extremely variable according to the chemistry of the polymer and the manufacturing process.

3.3 From particles to dispersion: latex stability

3.3.1 *The importance of controlling stability*

A latex is considered to be mechanically and chemically stable if it preserves the same dispersion state whatever stresses it may undergo. In other words, the latex particles must remain separate and in suspension in the aqueous phase throughout all the stages of its use.

1. During storage the latices must neither settle nor flocculate even over long storage times. The latex may also undergo large variations in climate conditions. It must be able to resist freezing temperatures and avoid forming skins or creams.
2. During shearing the dispersions used in the paper industry have high colloidal stability even under the harshest conditions, such as on modern blade-coating machines and metered size presses. On these kinds of machines the coating colour undergoes a double strain – intense shearing (close to 10^4–10^7 s^{-1} depending on machine speed) and high compression. If the latex is not mechanically stable enough, particle agglomeration will cause considerable trouble during coating color transfer (passage though filters or pumps) and in the paper machines.
3. During formulation the latices must be totally compatible with the different ingredients of the sizing and coating formulas. Otherwise, the rheologies may not be appropriate to the product's final use or the coating colors may be unstable during coating.

4. During paper coating, when paper is sized or coated on machine, substances from the base support can be released back into the coating color. These substances are generally cationic chemical compounds that can flocculate the latex and other anionic colloids in the colors. This may occur with alum for papers manufactured in acid medium, with polyacrylamide type cationic polymers, and with sizing agents for papers made in neutral medium.

If, however, the latex must have high colloidal stability to resist the mechanical strains and cationic chemical pollution inherent in the coating process, its stability must in no way disturb the physical chemistry of the wet end of the paper machine. If the coating color or wet broke is injected into the circuits of the paper machine the latex must be coagulated to keep polymer species from concentrating into sticky deposits known as white pitch. Coating latices are usually designed with this in mind. Used with conventional cationic agents present in the wet end of the paper machine, the latices flocculate completely and rapidly as very fine aggregates that are easily retained in the paper sheet as it is being made.

Nonetheless, as for any colloidal system, flocculation appears only at a given level of cationic agent. If the cationic agent is at too low a level there is no flocculation possible. This level is the cationic demand of the latex and corresponds to the amount of cationic species needed to cancel the zeta potential of the latex particle (Figure 3.2).

It is thus necessary to control the stability of the emulsions so as to get a

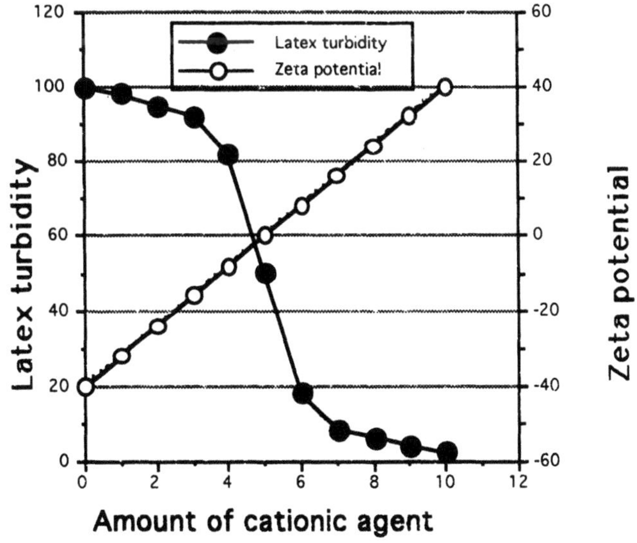

Figure 3.2 Latex flocculation behavior.

system that is stable during storage, during the formulation of the coating color and during paper coating, and yet sufficiently unstable so as not to disturb the physical chemistry of the wet end of the paper machine during recycling of the broke.

3.3.2 Stability control

Like any ultradivided system, a concentrated suspension of polymer particles, as described here, is thermodynamically unstable. These particles are in fact subjected both to Brownian movement and van der Waals type attraction forces. These forces become very strong when the distance between particles is small, and any collision between particles constitutes an inelastic shock. To forestall latex coagulation, the particles are protected by placing barriers around them to prevent them from getting so close to each other that van der Waals attraction forces become preponderant. In practice there are two types of protection: electrostatic and steric.

Electrostatic protection. To achieve electrostatic protection, electric charges are created on the surface of the particles. These charges induce the formation of an electrostatic potential barrier. The system is all the more stable as high energy is needed for the particles to cross this barrier, which keeps the particles from getting too close to each other and from falling to such a low energy state that the system coagulates.

These surface charges are created either by incorporating ionic emulsifiers into the medium or by using an ionogenic co-monomer during latex synthesis. This co-monomer is an unsaturated molecule with an ionizable cluster that slips into the polymer chains composed of the principal monomers.

The emulsifiers or/and the co-monomer then insert a hydrophilic grouping that preferentially arranges itself at the surface of the particle, thus generating surface electric charges if the exterior environmental conditions (pH, ionic strength, etc.) are favorable. Among these ionogenic co-monomers are principally found vinylic acids, such as acrylic acid, methacrylic acid, fumaric acid, maleic acid, itaconic acid, etc. (Figure 3.3(a)) and sulfonated groups derived from the action of the radical ion on the monomers (Figure 3.3(b)).

Steric protection. When hydrophilic macromolecules are fixed to the surface of latex particles, a diffuse layer is obtained which works as a mechanical barrier. The role and the mechanism of the protective diffuse layers is beginning to be better understood [4]. They can be achieved by adsorbing non-ionic emulsifers at the interfaces. These are molecules that are not ionized (as opposed to the previous case) but which comprise

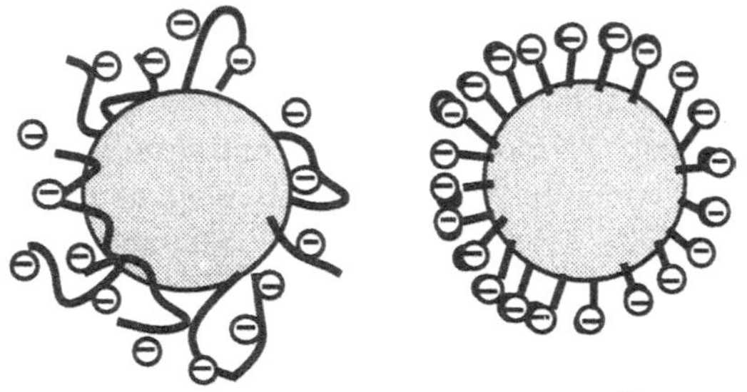

Figure 3.3 Electrostatic protection of latices by: (a) vinylic acids; and (b) sulfonated groups.

a highly hydrophobic part anchored strongly to the polymer particle and a long hydrophilic chain that develops in the aqueous phase. Steric stabilization has the advantage over electrostatic stabilization in that it is independent of such parameters as pH or ionic strength (i.e., ionizable salt concentration in the aqueous phase). These systems are, however, very sensitive to any disturbances capable of modifying the interactions between the aqueous phase and the polymer of the protective layer. Thus, if chemicals are introduced or temperature is raised, the system risks being destabilized. By judiciously adjusting the relative proportion of steric and electostatic protection, latices can be used with no major difficulty in the different stages of the paper coating process without becoming a disturbing factor in the physico-chemical equilibrium of the wet end of the paper machine.

3.4 The viscosity of latices and coating colors: the novel contribution of the particle state

3.4.1 Novelty of emulsion polymer latices

Emulsion polymerization makes it possible to obtain polymers with very high molecular masses (10^5 to 10^6 monomer units) and with viscosities of a few hundred centipoise, and the concentration can attain 50% by weight.

An emulsion latex is an aqueous medium of controllable viscosity independent of the nature and molecular characteristics (mass and structure) of the polymer, whereas the viscosity of a solution of hydrosoluble latex polymer depends on the nature, the concentration and the size of the molecules. There is no doubt that emulsion latices are the binders offering the best compromise between solids content and viscosity.

The polymers used have a density near 1 g ml^{-1}. This means the concentration by weight is close to the concentration by volume. Since the density of the most commonly used pigments, such as carbonate, is around 2.5, a latex with 50% active material (by volume or by weight) is as concentrated as a pigment slurry containing about 70% solids by weight. Latices are therefore among the most fluid materials, and the most heavily concentrated by volume, used in the paper industry.

Latices have little or no influence on the viscosity and the rheology of coating colors. It is thus necessary to formulate these colors with viscosity modifiers to obtain the rheology and water retention required for the color to be runnable on the coating machine. These viscosity modifiers can be natural, such as starch and guar gum; natural but modified, such as the carboxymethyl celluloses; or entirely synthetic, such as polyvinylic alcohols. There exist, however, some latices that provide both viscosity and water retention. These latices, better known as sole binders, are commonly used for high-speed on-machine coating. Starting from an alkaline pH, the sole binder determines the viscosity according to the application. They also have a significant impact on the rheology of the coating color as they can give appropriate pseudoplastic rheologies even with pigments reputed for their difficult rheologies.

3.4.2 The impact of the particle state and the chemistry of the latex on its viscosity

The viscosity of a latex obeys laws involving the concentration, size and interaction of the particles [5]. The rheology of the latex particles can be described by two fundamental effects in addition to the effect of the concentration of the dispersion: the electroviscous effects [6, 7].

The principal, or first-order, electroviscous effect, comes from a deformation of the ionic environment of the particle under the influence of shearing. There results an energy loss, which is responsible for the displacement resistance that increases the viscosity. To describe the viscosity of such media Smoluchowski has suggested extending the Mooney relations [8, 9], by including the electrostatic contribution in the intrinsic viscosity factor K (equal to 2.5 for spherical particles [10]), according to the following Mooney relation:

$$\eta = \eta^{\circ} \exp \left(\frac{K\Phi}{1 - \dfrac{\Phi}{\Phi_l}} \right)$$

In Smoluchowski's correction K becomes $K(1 + K_1)$, with

$$K_1 = \left(\frac{\Psi\varepsilon}{2}\right)^2 (\lambda_S \eta^\circ R^2)^{-1}$$

where

ε is the dielectric constant of the support fluid (water);
λ_S is the specific conductivity of the support fluid,
η° is the viscosity of the support fluid,
Ψ is the electrokinetic potential of the particle;
R is the radius of the particle;
Φ_l is the limit volume fraction that cannot be exceeded, corresponding to maximum packing of the particles.

From purely geometric arguments Φ_l is estimated to lie between 52% and 74%. The lower limit is calculated from a loose (simple cubic) and the upper limit from the tightest possible (face-centered cubic) packing of spheres of identical diameter.

This point is perfectly confirmed for latices (Figure 3.4). The viscosity of the polymer dispersions varies exponentially with the density of the surface charge, respecting the laws formulated by Smoluchowski, regardless of the internal chemistry of the latex particle.

The second-order electroviscous effect results from interaction between the double electric layers of the particles that pile up as the volume

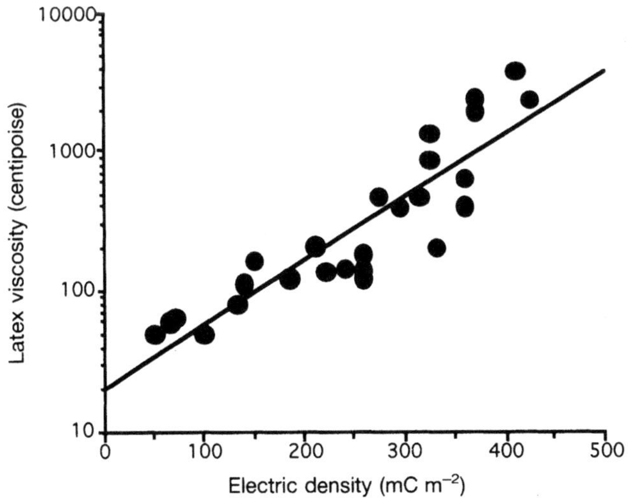

Figure 3.4 The viscosity of latex obeys the Smoluchowski relation, regardless of the internal chemistry of the latex particles.

fraction increases. This effect, which shows up as a high increase in viscosity, will be greater as particle charge rises, as particle diameter decreases, and as the ionic force of the medium drops. It is to be noted that this second-order electroviscous effect increases as the square of the concentration and appears only in extremely concentrated media.

3.5 From particle to film: latex coalescence

3.5.1 How to obtain a homogeneous film

Film formation is one of the most sought after properties of latices in paper applications. This property ensues from the dispersed form of the polymer. Film formation takes place when a liquid latex film dries (Figure 3.5(a)) [11]. The latex is first seen to concentrate, as the water penetrates into the support or gradually evaporates. The latex particles are packed in a hexagonal order, the most compact structure known in nature. At this stage the volume concentration of the latex reaches 74% (Figure 3.5(b)). When the system is sufficiently concentrated, the particles get close to each other and are progressively deformed. The latex particles then lose their spherical shape and take on the form of a rhombic dodecahedron, putting each latex particle in contact with twelve other particles (Figure 3.5(c)).

This is very similar to the structure observed in foams (Figure 3.5(d)). At this level of compactness the threshold of coalescence is attained. By coalescence is meant the fragmentation of the latex particle membranes, making it possible for the macromolecular chains of a particle to interdiffuse into the neighboring particles. This macromolecular diffusion destroys the periodicity and the arrangement of the latex particles to end up with a continuous, homogeneous polymer film (Figure 3.5(e)).

This phenomenon is triggered by capillary forces that arise when the system is very highly concentrated and there subsists an interstitial film of water around and between the latex particles.

These forces are linked with the water–air interface tension, the shape of the particles and the radii of the interface curves. If the polymer is flexible at its formation temperature the particles, distorted by the pressure of the capillary forces, coalesce. Depending on conditions, the coalescence can go as far as the interdiffusion of the macromolecules of the core (Figure 3.5(e)).

In practice coalescence is possible if the latex particles are in the presence of water and there is a minimal temperature ensuring particle deformation. This critical temperature is the minimal film formation temperature (MFFT). The paper industry uses polymers whose MFFT is between −30°C and −40°C.

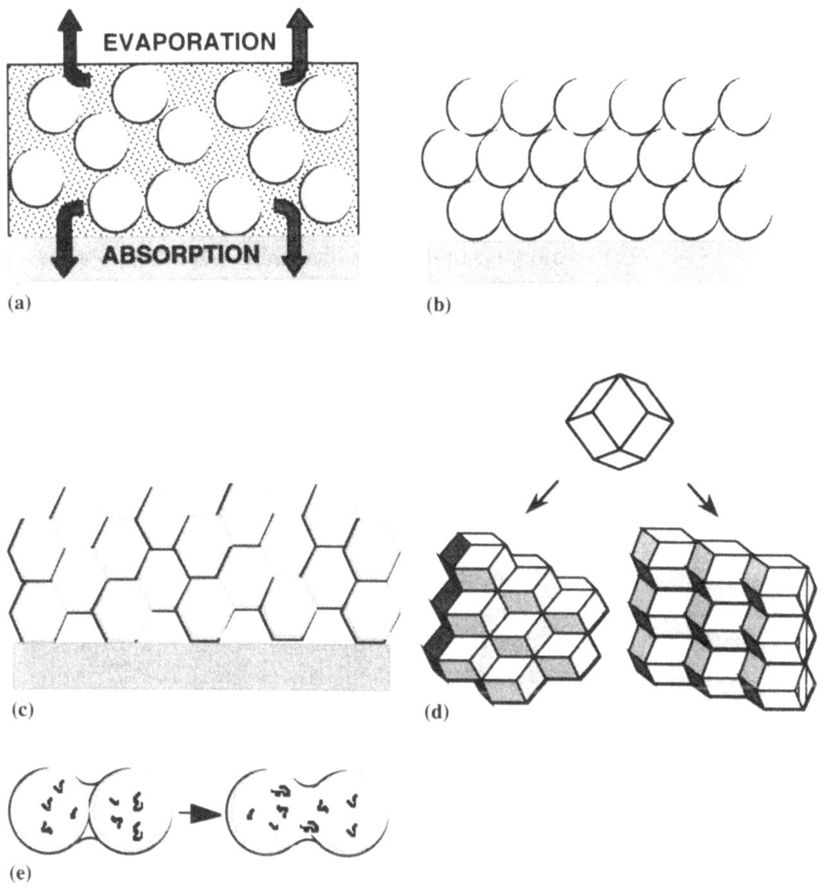

Figure 3.5 Latex consolidation and film formation: (a) film formation by the drying of the liquid latex film; (b) latex particles in a hexagonal packed order; (c) deformation of hexagonal packed particles into a rhombic dodecahedron form at higher concentrations; (d) foam structure (for comparison with structure shown in part (c)); and (e) the creation of a continuous homogeneous polymer film by coalescence.

The use of a continuous polymer film on the surface of paper is common in the paper industry. We can cite, for example, machine-coated buffered-surface paper. In this application the polymer provides a physical barrier to the penetration of fluid. This fluid can be water or oil. To a lesser extent this barrier effect, through latex film formation, is used to improve paper sizing. This physical barrier, as measured by the Cobb value obtained, is far more efficient than bulk sizing or starch sizing. By using latex instead of starch, sizing formulae with higher solids content can be obtained which reduces the penetration of the sizing layer into the base support.

3.5.2 Wetting latex particles on pigments

Latices are not always used at a concentration that allows homogeneous film formation, i.e., generally above 25 parts by weight, depending on the nature of the pigment used. This is the situation for coated papers used in offset and rotogravure printing, where the latex concentration lies between 5–15 parts by weight. Here it is important to observe the behavior of the latex during the formation of the coating, i.e., in the presence of mineral pigment. This observation is made possible by the technique of atomic force microscopy (AFM) [12].

In this technique a sharp diamond tip mounted on a cantilever sweeps over the surface of the sample on an *x–y* plane. This tip follows the topography of the sample, causing the cantilever to be deflected vertically (*z* axis). By measuring the variations in this deflection as a function of the *x–y* scan, images of the unevenness of the surface are reproduced. The underlying principle of the technique rests on the interaction forces that arise between the last atoms of the tips and the surface atoms of the sample surface (Figure 3.6).

A typical AFM image is shown in Figure 3.7. This image shows that the latex particles are well dispersed at the surface of the calcium carbonate crystal and that they do not cover the whole surface of the crystal, given this particular latex content. Some of the particles are in contact with each other and have coalesced as described above. It is to be noted that as the system has dried, the latex, instead of flocculating, has welded to the shape of the pigment by wetting its surface.

The diameter and the shape of the latex particles can be determined by this technique. It shows that starting with a latex whose initial diameter is 120 nm, latex beads 260 nm in diameter and 30 nm high that are well dispersed on the crystal surface can be obtained. These measurements demonstrate clearly that the latex wets the crystal surface by spreading. By

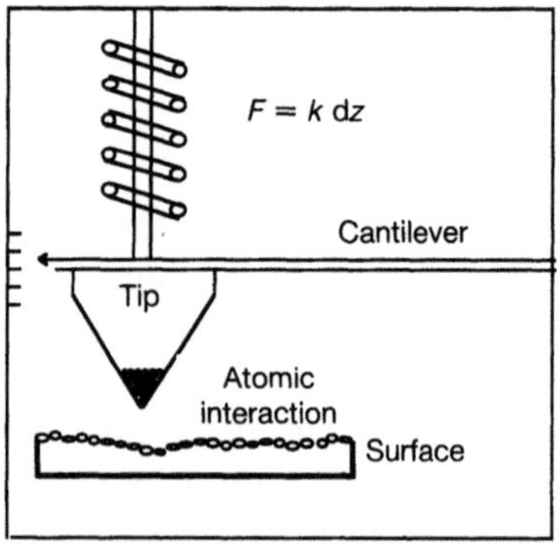

Figure 3.6 Schematic diagram of the principle behind the atomic force microscopy technique. F = interaction force; k = cantilever force constant; z = coordinate direction.

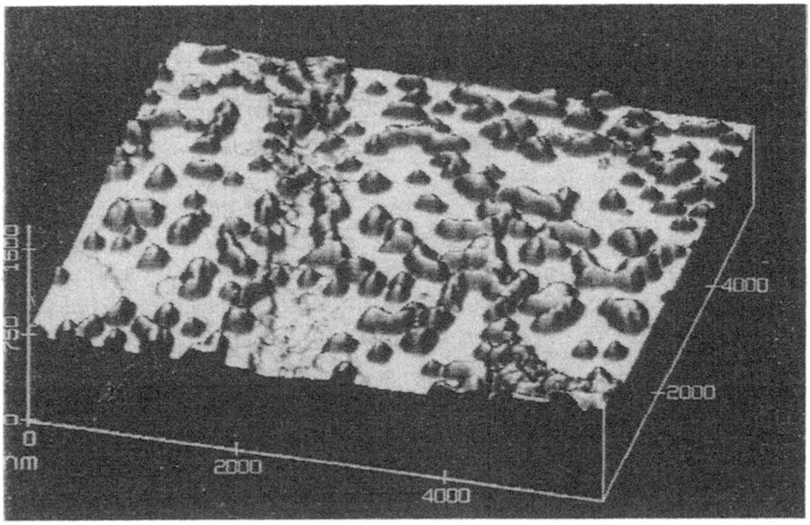

Figure 3.7 An atomic force microscopy constant-force image of latex particles placed on a calcium carbonate crystal.

observing latices of variable rigidities the latex is seen to be able to spread only if the polymer is soft at the system's formation temperature. If the polymer is rigid at formation temperature, the latex particle remains spherical, which prevents it from spreading on the crystal surface.

There is therefore a critical temperature beyond which the latex particle spreads over the pigment. This critical temperature is the glass transition temperature (T_g). The paper industry habitually uses polymers with a T_g between −30°C and 40°C.

3.5.3 *Preventing film formation*

Film formation is not always desirable. With plastic pigments it must even be prevented. If a polystyrene latex is used, the film formation temperature is so high (between 100°C and 110°C) that the latex cannot form a film under normal manufacturing conditions in the paper factories. As this latex cannot form a film or wet pigments, it has no binding power and thus has properties like those of a pigment, whence it is called a plastic pigment.

The benefits gained by using plastic pigments are:

- a gain in gloss, with the same value for any particle size;
- a gain in brightness and opacity, with an optimum at 0.4 microns;
- increased ink and air porosity, with a maximum between 0.4 microns and 0.5 microns;
- increase in printed gloss and decreased roughness as particle size increases.

The characteristics of brightness and opacity are contingent on the porosity of the coating. The porosity, in turn, depends on the fit of the shapes and particle size distributions of the pigments that are present together – kaolin (hexagonal platelets), carbonate (nearly spherical), and plastic pigments (spheres). The degree of fit brings about more or less disorder when the coating is applied to the base support.

The spheres of plastic pigment stick in the grooves of the coatings, evening up the surface. This effect is amplified when the plastic pigment beads are crushed during calendering. Plastic pigments thus help reduce the roughness of the surface and boost the unprinted gloss of the paper. This improvement in unprinted gloss of course makes for higher quality printed gloss.

Although use of plastic pigments degrades the cohesion of the coating, the cohesion is maintained by increasing the proportion of latex by 1 part per 10 parts of plastic pigment. The need for more binder when using plastic pigments is directly linked to their higher specific surface area, which is a result of their low granulometry.

3.6 From latex to coated paper: the role of polymers in a paper coating

3.6.1 *Binding power*

This remarkable ability of latices to form a film and wet the surface of pigments partly explains their binding power, a latex property heavily exploited by the paper industry for many years.

Not all latices develop the same binding power. This property is linked to the ability of the latex to wet the surface of pigment and thus, as we have just seen, to the polymer's glass transition temperature. There is therefore a direct relationship between the glass transition temperature, T_g, of the latex polymer and the dry pick resistance of the coating.

By analyzing the mechanism of pick, a certain number of hypotheses can be iterated.

- Inking porosity of the coating: although coated papers exhibit a good correlation between inking porosity and pick, this parameter is not relevant. As evidence of this, it suffices to recall that calendering the coating profoundly modifies ink setting without affecting pick at all.
- The elasticity of composites: polymer elasticity affects the elasticity of the coating. Coating cohesion is thus contingent on the rigidity of the latex used. The more rigid the latex, the weaker the dry pick resistance.
- Adhesion of latex to the pigments: the stronger this adhesion, the greater the binding power of the latex. Adhesion depends on the

chemistry of the polymer, which determines the ability of the latex to wet the pigment particles.

- Peeling of the coating from the base support: pick can be represented as a peeling of the coating from the support at the interface. In this case a composite adheres to the base paper by an ultrathin film of latex, probably close to a single monolayer of particles.

Generally speaking, although the adhesive power of the latex seen from the point of view of its ability to wet the surface of pigments is essential to binding power, the elasticity of the polymer is not a negligible factor.

3.6.2 The latex polymer: an elastic material

In its paper application, a latex loses its particle shape to take on the attributes of a film or to be strongly welded to pigment particles. The properties thus obtained are related to the elastic behavior of the polymer, that is, to its Young's modulus.

The Young's modulus of certain materials varies with the temperature. This change may be continuous, or it may be discontinuous if the material goes through a phase change. An example of the latter case is a latex that can be rigid, rubbery, or fluid at different temperatures.

The temperature at which a polymer passes from a glassy state to an elastic state is the glass transition temperature, T_g. The variation in the Young's modulus as a function of temperature can be measured with precision by using the techniques of dynamic mechanical analysis (DMA), on condition that the amplitude of the imposed periodic deformation is small enough to remain linear.

The results are presented as variations of Young's modulus E', as a function of the frequency, or, if the frequency is fixed, as a function of temperature. A typical spectrum for a polymer with a value of T_g around 20°C is given in Figure 3.8.

On this spectrum, three distinct ranges of temperature appear.

- At low temperatures the polymer has a high Young's modulus, E', that is virtually independent of the temperature. This state is characteristic of the glassy state of the polymer.
- At around 20°C the Young's modulus suddenly drops by about three powers of 10. These variations express the phase change of the polymer as it passes from the glassy state to the elastic state.
- Beyond the glass transition the Young's modulus varies little, with a negative or zero slope. Values of E' between 10^6 MPa and 10^4 MPa are characteristic of the polymer's rubbery state.

After this plateau, at higher temperatures, there is a final zone of transition to a fluid state. This fluid has its own peculiar rheological properties, described by the pseudoplasticity index (n) and by the consistency (K).

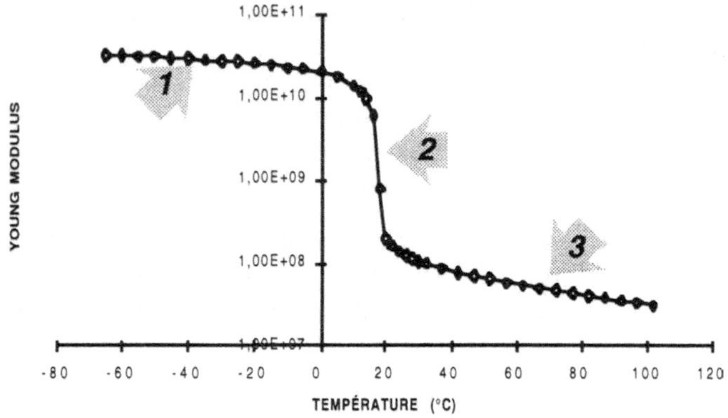

Figure 3.8 Dynamic mechanical analysis spectrum of latex.

These two parameters depend on the macromolecular structure of the polymer and describe the propensity of the polymer to be deformed under pressure and at very high temperature. In particular, for paper to be used in roto-offset printing, the blistering resistance is directly correlated with the rheology of the polymer at high temperatures and seems not to be related to the polymer's permeability to steam [13].

The elastic modulus of the polymer plays an essential role in paper stiffness. As paper stiffness is a constant requirement in sheet offset and roto-offset printing, it is essential to integrate the role of the latex used.

The coated paper is not homogeneous; its Young's modulus differs between the paper core (base paper) and the coating. The rigidity of a coated paper is given by the following equation:

$$R = \sum_{i} E_i I_i = \sum_{i} E_i \left(h_i x_i^2 + \frac{h_i^3}{12} \right), \qquad (3.1)$$

where

E_i is the Young's modulus of layer i;
I_i is the moment of inertia of layer i with respect to the bending plane;
x_i is the distance from the center of layer i to the bending plane;
h_i is the thickness of layer i.

Thus, for symmetrically coated papers:

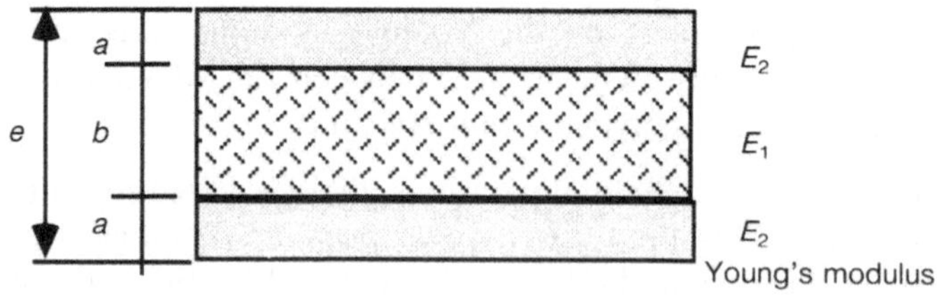

If we look at a paper support with Young's modulus E_1 and thickness b on which two similar coating layers, of thickness a and Young's modulus E_2 have been deposited, the rigidity, R, of this coated paper is given by:

$$R = E_1 \frac{b^3}{12} + 2E_2\left(a\left(\frac{a+b}{2}\right)^2 + \frac{a^3}{12}\right). \tag{3.2}$$

When a layer of coating is applied to a support, the paper stiffness is changed by $R°$ according to the expression:

$$R - R° = \Delta R \simeq E_2 \frac{b^2 a}{2}, \tag{3.3}$$

where as before a is the thickness of the coating layer (a function of its coat weight) and E_2 is its Young's modulus. Therefore, owing to increased thickness, a coated paper's bending stiffness is always greater than that of an uncoated support. This difference in stiffness increases as the Young's modulus of the coating increases.

A high Young's modulus is obtained by using a rigid binder in the coating color. This notion of latex rigidity mirrors the notion of glass transition temperature. The higher the value of T_g, the more rigid the polymer, regardless of the chemical nature of the latex (Figure 3.9). Thus it is seen that the elastic properties of latices can be used to tailor the stiffness of coated papers.

3.6.3 Polymer chemistry as a modifier of paper surface energy

The surface energy of a coated paper is an important parameter in determining the spread and capillary penetration of liquids. In practice these two properties affect the following qualities:

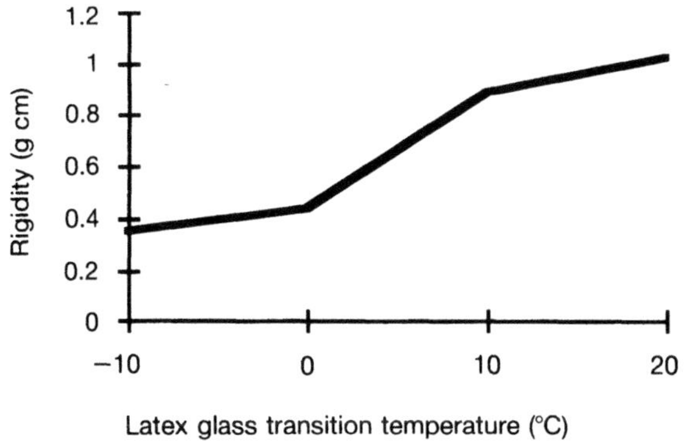

Figure 3.9 The influence of latex rigidity on paper stiffness.

- ink refusal, determined by the ability of the paper to absorb offset fountain solution;
- ink intensity and print dot size, especially in liquid ink processes;
- edge binding and padding with hydrophilic glues, determined by the wetting of these glues to the paper [14].

To find out if a liquid will spread on a paper surface, the polar and dispersive components of the liquids and the paper have to be known, so as to apply Young's rules. In practice the wetting energies of papers depend on the nature of the pigments used and the chemistry of the latex.

Generally speaking, acetete-based latices give highly hydrophilic coatings; styrene–butadiene-based latices give moderately hydrophilic coatings; and acrylic-based latices give rather hydrophobic coatings.

3.6.4 Porosity and ink setting in coated papers

Ink setting in coated papers depends mainly on the porous volume of the coating. It is clear that both a rise in the latex level and in the calendering process itself diminish the porous volume of the coating. The loss of porous volume during calendering results from a crushing of the porous structure under the strain of the calender. The softness of the latex determines the degree of crushing. The softer the latex, the greater the effect of the calendering. For coated papers with different latex levels, the variation in the porous volume obtained will depend on the latex level. The higher the latex level, the lower the porosity.

In practice the porous volume decreases regularly as the level of latex increases to 25 parts. At rates of 25 parts and above the coating no longer has any porosity. This behavior is well known; the value of 25 parts corresponds to the critical pigment volume concentration. The polymer then exactly fills the voids between the pigment particles and film formation is total (Section 3.5.1).

The impact of the porous volume on industrial ink setting can be studied by measuring the ink set off index. Figure 3.10 shows the optical density as a function of the level of latex introduced into the formulation of the coating color for an ink contact time of 90 s.

There are two different areas of behavior: the behavior of coatings containing from 5 parts to 25 parts of latex, and that of the coatings formulated with more than 25 parts of latex.

In the first part of the curve the optical density rises with the level of latex introduced in the formulation, which implies a declining level of ink absorption by the coating. Under these conditions it is essentially the coating porosity that determines ink setting. The fewer latex particles present, the more porous the coat, and the more ink that is set.

In the second part of the curve, where the colors are formulated with

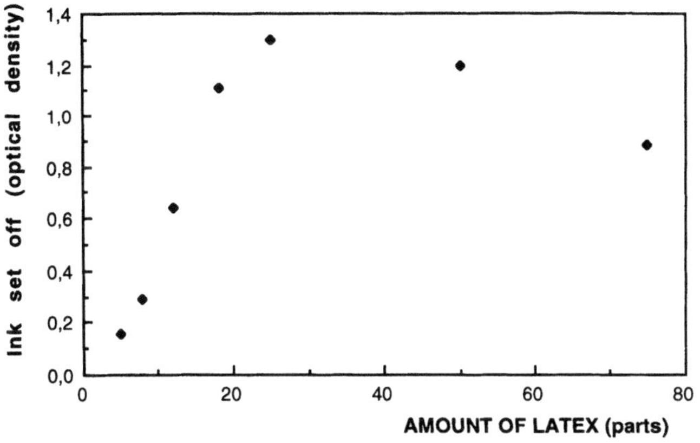

Figure 3.10 The impact of the amount of latex in the formulation on the penetration speed of Brancher industrial ink.

more than 25 parts of binder, the optical density of the ink set off drops as the level of latex in the coating increases. This behavior indicates an increase in ink absorption. Since there is no longer any physical porosity beyond 25 parts of latex, ink absorption is affected only by the polymer, which, depending on its internal chemistry, can swell in the presence of solvents and ink thinners.

There is thus a minimum of ink absorption corresponding to a coating containing 25 parts of latex. This represents the transition point between the two modes of ink absorption: porous structure and polymer swelling.

When the coatings are formulated with levels of latex below 25 parts, ink penetration into the porous structure constitutes the overriding mechanism that fixes the ink to the coating. Certain colors for industrial paper coating have lower levels of binder. Under these conditions the nature of the latex, because of its affinity with the liquid phase of the ink, has little effect on the paper's ink setting capacity.

3.7 Conclusions

The major properties of latices are linked to their stability, their particle state and their ability to form a continuous film or to adhere to pigments. These four points are intrinsic in the latex insofar as they depend on the dispersion state of the polymer particles.

The chemistry of the particles, far from being unimportant, represents the added value of the latex. A great many of the properties of a paper depend on it: binding power, stiffness, blistering resistance. As the latex chemistry alters the hydrophilic–lipophilic balance of the surface of the paper, it also affects wettability, for instance, performance under hot melt or aqueous gluing, ink refusal in offset printing or receptivity to fluid inks.

In summary, latex plays an essential role in paper properties. This is all the more important in that the chemistry used can be tailored to demand and is restricted only by the technological limits of emulsion polymerization.

References

1. Isoard, J.-C. (1988) Synthetic latexes: binders for the paper industry. 21st Annual Congress of Pulp and Paper, Brazil, November.
2. Larraillet, P., Delaplace, P. and Isoard, J.-C. (1985) Les liants de couchage. *Revue ATIP*, February.
3. Daniel, J.C. (1988) Les latex synthétiques. *Pour la Science*, March.
4. de Gennes, P.G. (1987) La matière ultradivisée. *Pour la Science*, number 111 (January).
5. Bacquet, G. (1995) Physico-chimie des bains de couchage, *Revue ATIP*.
6. Krieger, I.M. (1972) *Advan. Colloid Interface Sci.* **3**, 111–36.
7. Conway, M. (1960) *Rheology*, vol. 31 (ed. F.R. Eirich), Academic Press, New York, pp. 83–121.
8. Mooney, M. (1951) *J. Colloid Science*, **6**, 162–70.
9. Mooney, M. (1957) *J. Colloid Science*, **12**, 243.
10. Einstein, A. (1906), *Ann. Physik*, **14**, (19), 289.
11. Joanicot, M., Cabane, B. and Wong, K. (1993) Structure of latex film. 1993 TAPPI Coating Conference, PG 175/185.
12. Granier, V., Sartre, A. and Joanicot, M. (1992) An atomic force microscopy study of the wetting of an inorganic surface by latex particles. EURADH, September.
13. Schwob, J.M. and Guyot, C. (1991) The effect of latex/polymer properties on blister resistance of coated paper. TAPPI Conference, May.
14. Dobler, F. (1995) Spezifische anforderungen an das streichen von selbstdurchschreibe-papieren. PTS Symposium.

4 Starches

N.-O. BERGH

4.1 Introduction

Starch is the most used additive derived from renewable raw material in the paper industry. Of the 2.8 million tonnes of starch used in industry within the European Union in 1995, more than 46% – 1.3 million tonnes – was consumed by the biggest customer group of the starch industry: the paper industry. The consumption of starch has almost doubled in the past 10 years. This increase is not just the result of the growth of the paper industry, but also the increased use of starch in the different paper grades. Some 1.8 kg of starch were used on average per tonne of paper in 1995. This statistical average value, specific starch demand, also covers paper grades where no starch is used. Thus, some paper grades consumed as much as 50–70 kg starch per tonne. After fibres and pigments, starch is the third most important material for the paper industry as regards quantity. This starch use is apportioned as follows:

- surface application, 68%;
- wet-end addition, 16%;
- coating, 11%;
- spray, 5%.

The starch used in the paper industry is derived from different plants. Potato, maize and wheat are the most frequently used starch sources in Europe, though small quantities come from tapioca, barley and waxy maize (see Figure 4.1).

The growth of the paper industry's starch use in recent years has been driven by economic as well as technical reasons.

4.2 Principles of starch technology

4.2.1 Chemical principles

All higher plants containing chlorophyll form carbohydrates such as starch and cellulose as deposit and skeleton substances from carbon dioxide, water and light energy.

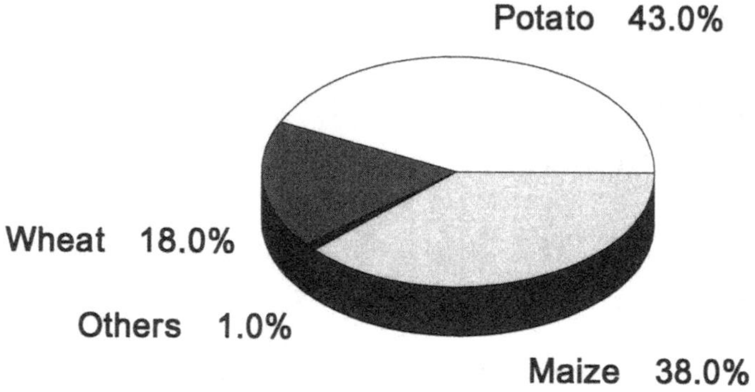

Figure 4.1 Starch consumption in the paper industry (raw materials).

In this chemical process – without doubt one of the most significant overall, but which is still not yet completely understood – long-chain molecules are produced from the initially formed D-glucose by way of enzymatic polymerization. These chain molecules may be either starch or cellulose, depending on the method of molecular linkage.

In the case of water-insoluble cellulose, the individual glucose units are linked by means of so-called β-glucosidic bonds, while the same units are linked by α-glucosidic bonds in water-soluble starch. Identical monomer units linked together to form different structures of the natural polymer produce completely different chemical and physical properties (Figure 4.2).

Starch is stored mainly in the roots and seeds of plants in the form of granules. The granules possess a characteristic morphology and granulometry for the respective plant types.

Potato starch consists of relatively large oval granules, whereas maize starch granules are smaller and round or polygonal. Wheat starch granules have a bimodal size distribution, so that the large grains account for only 20% of the total number but for 90% of the total weight (Figure 4.3).

If starch granules are observed in an optical microscope under polarized light, it is possible to see characteristic crosses as the result of birefringence. The centre of each cross indicates the growth centre.

All starches consist of two types of molecules, namely amylose and amylopectin. Amylose is made up of a chain of 600–3000 glucose units linked by α-1.4 bonds and fixed in a helical structure by intramolecular hydrogen bonding. The amylopectin molecule is branched by additional α-1.6 bonds (Figure 4.4).

The proportions of amylose and amylopectin vary, depending on the type of plant. Genetic selection makes it possible to breed types where these polymer fractions exist in a specific ratio. One of the most important

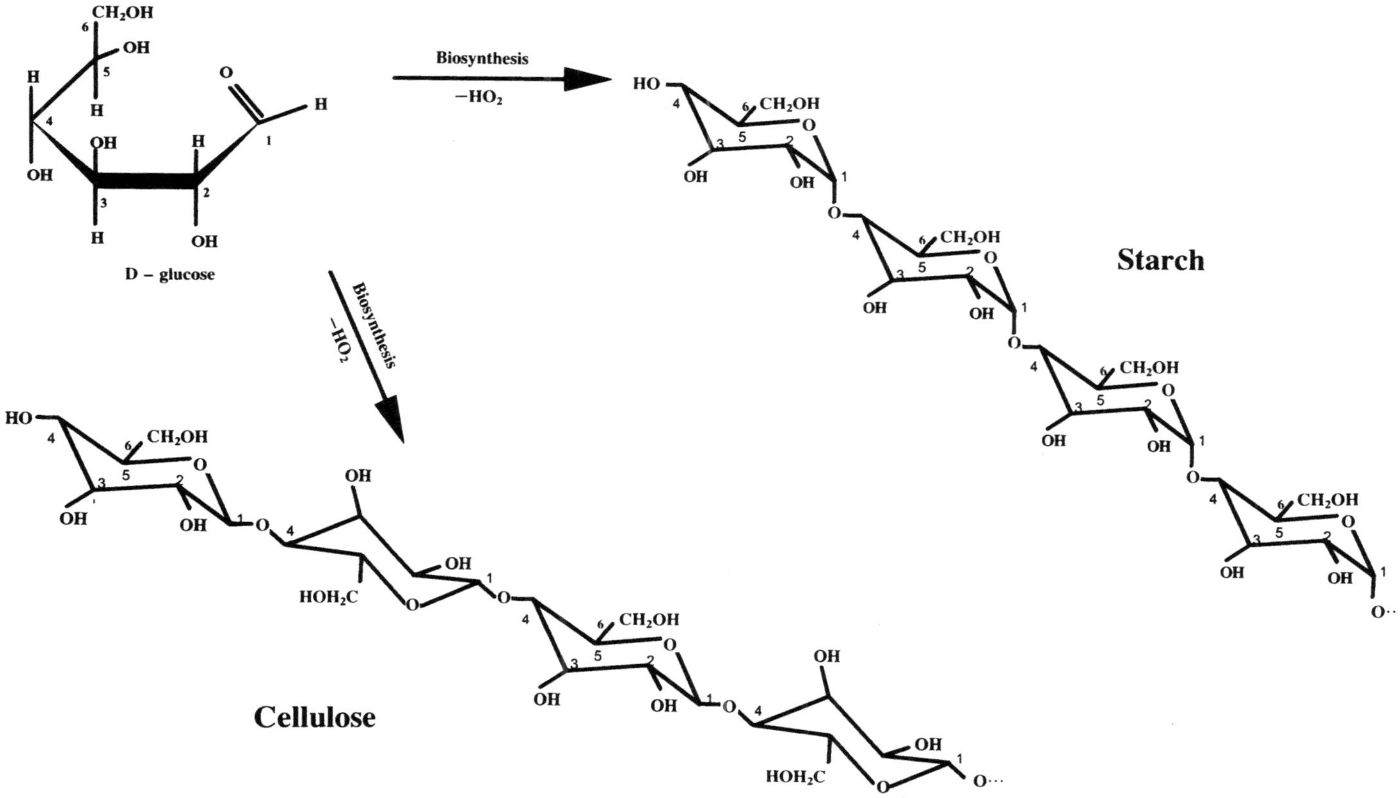

Figure 4.2 Biosynthesis of starch and cellulose.

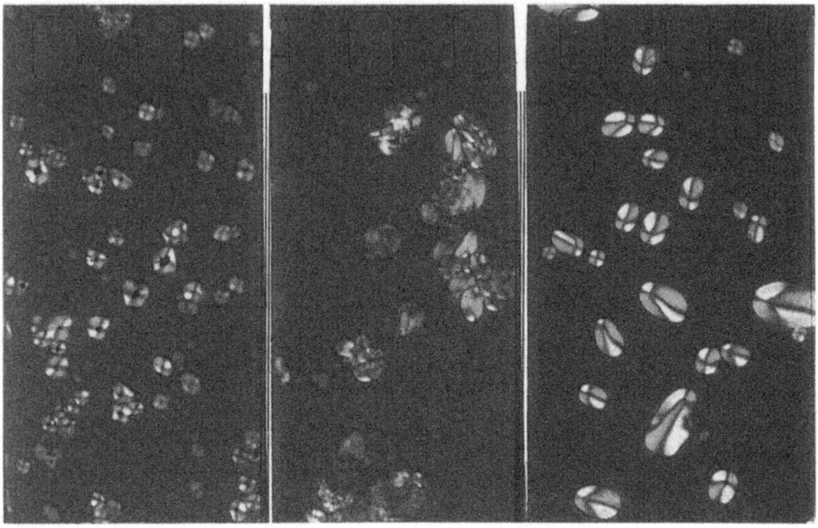

Figure 4.3 Microscope pictures of various starch types in polarized light. The centre of each cross is a growth centre.

breeds of this kind is waxy maize starch, which contains 99% amylopectin (Table 4.1). However, the quantity of starch in the various plant types also differs greatly, depending on whether it is obtained from roots (tapioca), tubers (potatoes) or seeds (wheat, maize).

It is of interest to note that the production of one tonne of starch requires the following growing areas and quantities of raw material:

- potato: 0.14 ha, 5 tonnes of raw material;
- maize: 0.24 ha, 1.6 tonnes of raw material;
- wheat: 0.33 ha, 2 tonnes of raw material.

4.2.2 Physical properties of starch

The moisture content or equilibrium moisture content under normal conditions (20°C, 65% relative humidity) is 13% for maize and wheat starch and 19% for potato starch.

Native starch is insoluble in cold water. Owing to their higher density, the starch granules are precipitated as soon as a suspension is no longer stirred. When the concentration of a starch slurry reaches 45% or more, it becomes dilatant, that is, its viscosity increases so rapidly with shear stress that it can no longer be stirred or pumped.

If the starch suspension is heated, the granules begin to swell at a temperature which is characteristic of each starch type. The starch granules swell to many times their original volume and lose their typical granular

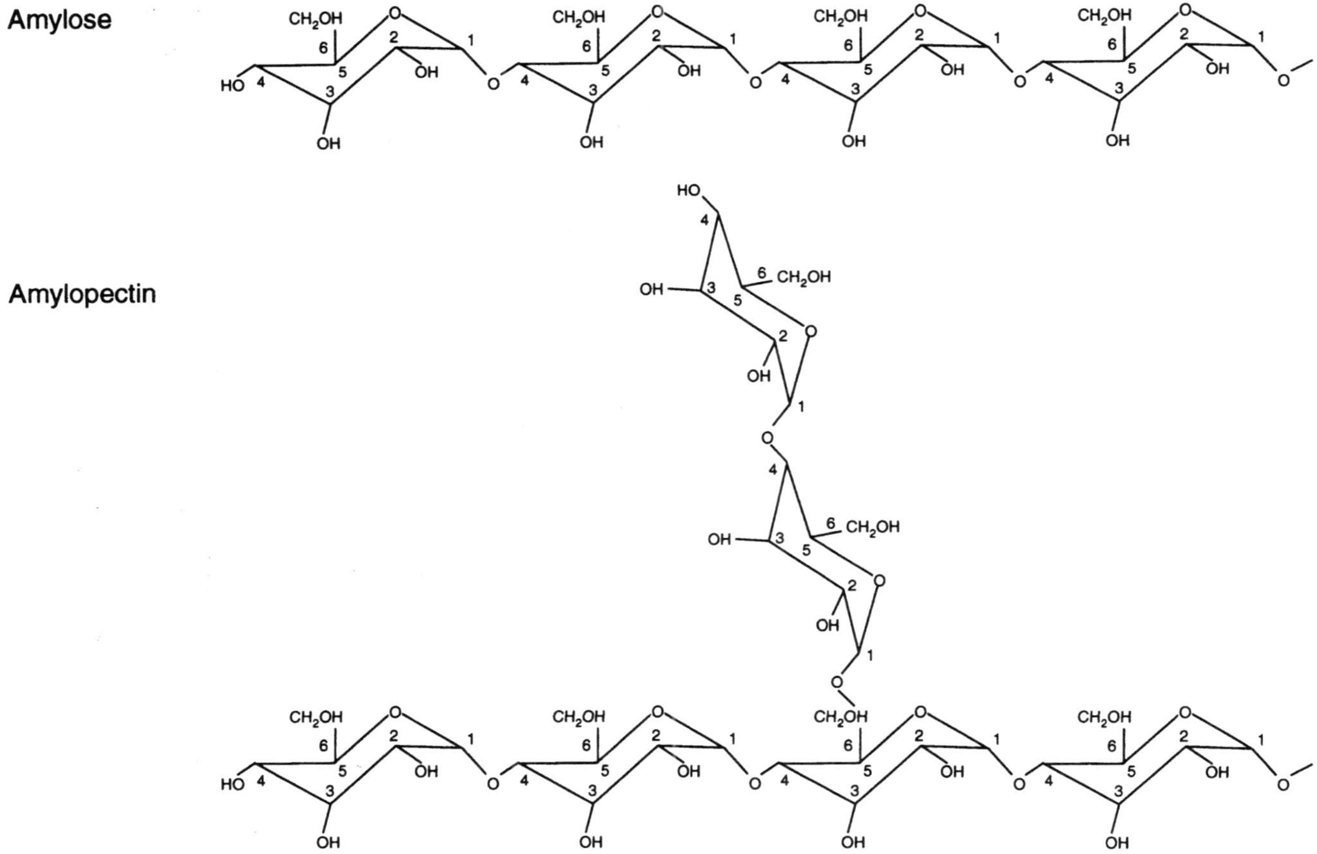

Figure 4.4 Polymer fractions of starch.

Table 4.1 Starch parameters

Parameter	Maize	Wheat	Waxy maize	Potato	Tapioca
Moisture (%)	13	13	13	18	13
Particle size (μm)	5–25	15–40	3–26	15–100	5–25
Gelatinization temperature (°C)	67	61	68	63	65
Amylopectin (% starch)	75	78	99	77	80
Proteins (% hydrogen bonded)	< 0.4	< 0.3	< 0.5	Traces	0.1
Lipids (% hydrogen bonded)	0.2	0.2	0.2	0.0	0.0
Phosphates	< 0.1	< 0.2	< 0.1	0.3	< 0.1

structure. At the same time, the characteristic birefringence of the granules disappears as swelling progresses. This process is referred to as gelatinization.

The amylose fraction of starch never completely dissolves, in the strict chemical or physical sense, even after boiling and gelatinization. Additionally, it displays a tendency to create hydrogen bonds and reorganize itself into linear bundles after a time.

This effect is called retrogradation, a process which results in the boiled aqueous solution becoming opaque. The amylose can be precipitated if the solution possesses low viscosity and is stored for a long period of time. In the case of solutions with higher viscosity, rigid gels may be formed from which water can be expelled in a reversal of hydration. This process is called syneresis and is irreversible in most cases (Figure 4.5).

In addition to these changes it should be noted that in maize and wheat starch paste, the amylose sometimes forms crystalline complexes which may be precipitated in the form of small particles under specific conditions: long storage time; high temperature (above about 65°C); and low pH (< 7). This amylose precipitation takes place if the starch–lipid complexes are broken up by cooking temperatures above 120°C. In the case of potato starch, which contains no lipids, this amylose precipitation can be induced by the presence of surface active substances alone.

These amylose particles look almost like ungelatinized starch granules under polarized light, but are larger and display concentric rings in addition to the polarization crosses (Figure 4.6).

Amylopectin is more resistant to this aggregation effect owing to its branched structure. Correspondingly, its retrogradation tendency after boiling is less marked. Amylopectin pastes retain their flow properties over a long period of time and any signs of slight retrogradation disappear after renewed heating.

Table 4.1 lists the properties of regular starches important in paper manufacturing.

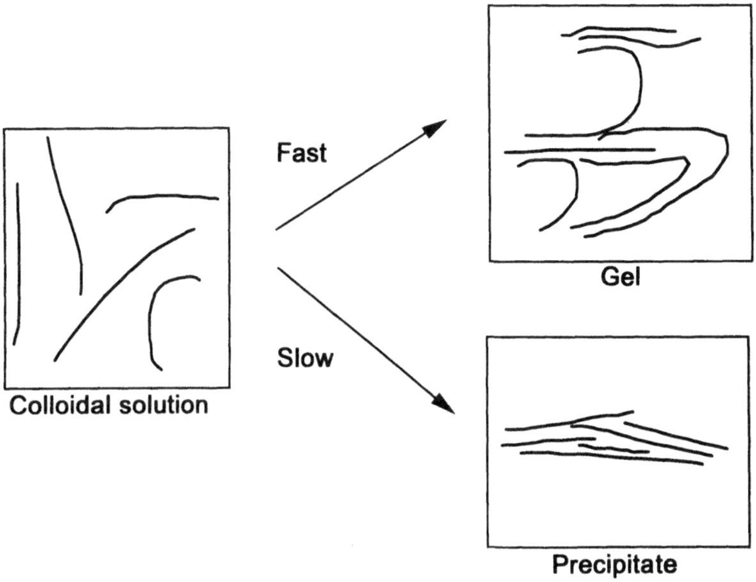

Figure 4.5 Retrogradation of amylose.

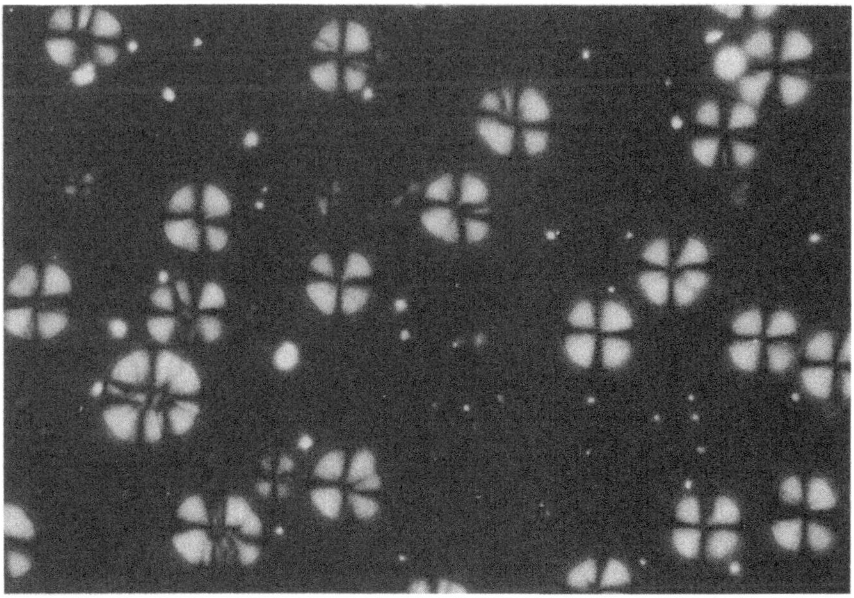

Figure 4.6 Retrograded amylose in polarized light.

4.3 Starch modification

Although pure native starch is used in the paper industry, it meets the physical demands of modern industrial processing only to a restricted extent. In many cases, the starch properties, such as binding or adhesive strength, water retention capacity, rheology or film formation properties, first have to be adapted to the special requirements of an application by physical or chemical modification.

4.3.1 Physical modification

It is possible to destroy the crystalline structure of starch granules by the influence of heat and mechanical shear forces, obtaining an amorphous material which dissolves easily in water to give a colloidal starch solution. Such starches are usually produced by roll drying or extrusion.

As a result of this additional treatment and the high amount of energy required to dry the hydrated starch, these products can be used economically only where starch cooking at the point of application is not possible.

4.3.2 Chemical modification

Chemical modification of starch can be performed so that:

- only the viscosity is reduced – degradation;
- chemical groups are introduced into the starch molecule – substitution;
- starch molecules are linked to each other – cross-linking.

It is also possible to combine any of these modification processes.

Degradation. The viscosity of starches can be reduced by various degradation methods: acidic, oxidative, enzymatic or thermal. Thermochemical degradation as used in the paper industry can be considered either as oxidation or as an acidification, depending on the degradation conditions.

Acid modification. If a starch suspension is treated with acid below its gelatinization temperature, partial degradation of the starch molecules results. However, the external form of the starch granules and their birefringence properties are not visibly influenced.

In this way, starches are obtained which form pastes with low hot viscosity at the same concentration. The fact that shorter chains are present facilitates macromolecular reorganization and the retrogradation tendency of the pastes is increased.

Owing to their high retrogradation tendency (Figure 4.7), these starches

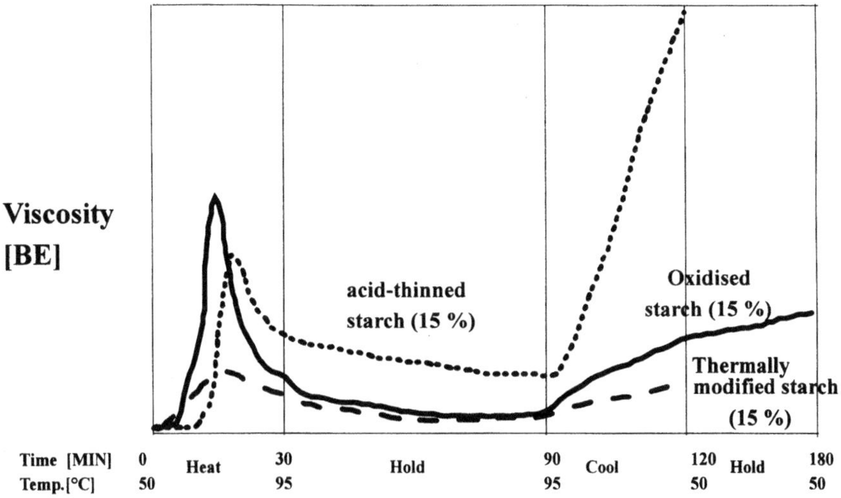

Figure 4.7 Brabender curves of different starches. Viscosity is shown in Brabender units.

find only restricted use in the paper industry, though they are significant intermediate products in the manufacture of starch ethers and esters with different viscosities.

Enzymatic and thermochemical conversion. Controlled reduction of the molecular weight and thus of paste viscosity can be achieved by enzymatic or thermochemical degradation of the starch molecules as well as by acid hydrolysis. Thermochemical degradation is performed by high-temperature cooking at 130°C–150°C with the addition of an oxidant such as potassium persulphate or ammonium persulphate, or hydrogen peroxide. In this process, chemical conversion does not take place in the suspended granule below the gelatinization temperature but in the already dispersed cooked paste. These degradation steps are therefore not performed in the starch factory but in the paper mill. The addition level of oxidants can vary from 0.1%–1.5%.

The differences in the behaviour of starches depolymerized by chemically very similar mechanisms are thus of a primarily physical nature. This is evident in the lower retrogradation tendency of oxidant-degraded starch pastes, compared with acid degradation. This results because oxidative degradation occurs in solution or in the pasted condition, whereas in acid hydrolysis it occurs by ring-splitting, which increases the potential for hydrogen bonding by developing new carboxyl and carbonyl groups.

An enzymatically degraded starch paste differs from a thermochemically degraded starch paste by its lower retrogradation tendency. Such differences are seen only at higher concentrations – as Figure 4.8 shows. The

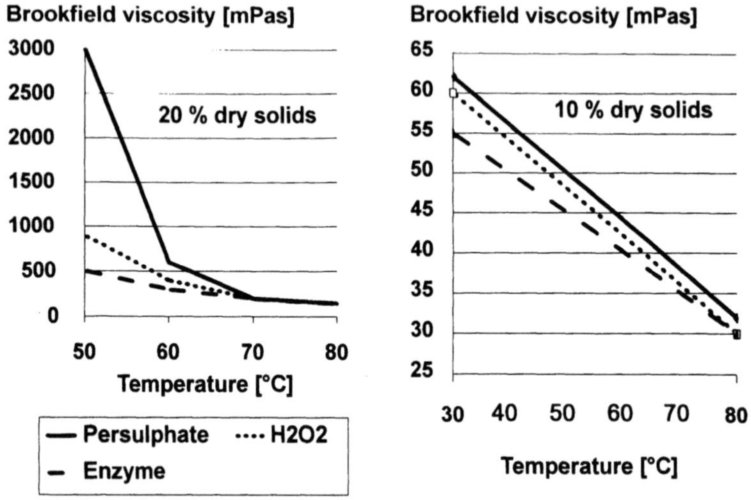

Figure 4.8 Retrogradation behaviour of thermochemically and enzymatically degraded starches upon cooling of the starch pastes.

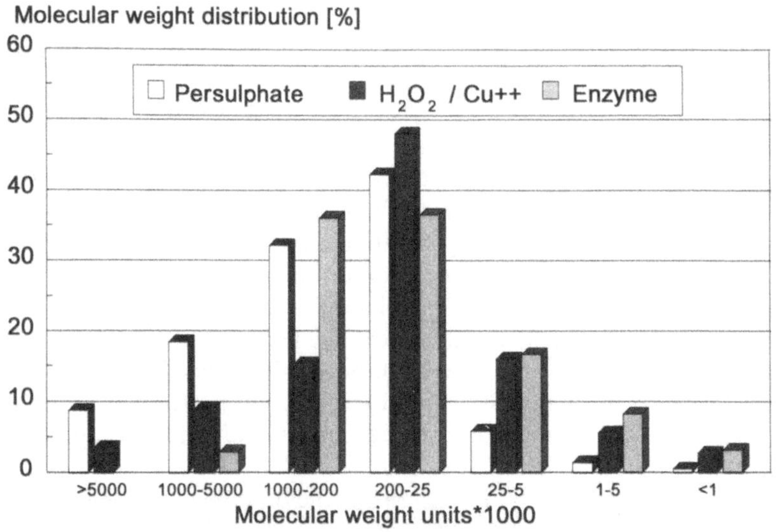

Figure 4.9 Molecular weight distributions of starches as a function of the degradation method.

improved visco-stability of enzymatically degraded starch is a result of the higher content of low molecular weight starch fractions.

The different levels of low and high molecular weight starch are shown by the molecular weight distribution curves. Figure 4.9 shows that higher contents of low molecular weight starch are obtained by oxidative degradation with hydrogen peroxide and with enzymatic degradation than with persulphate treatment.

Substitution

Oxidized starch. Native, granular starches can be depolymerized by the starch producer by treatment with one of several oxidants to give low-viscosity starch pastes with a high solids content and high viscosity stability (Figure 4.7).

In the production of standard commercial oxidized starch, an alkaline hypochlorite solution containing approximately 5% active chlorine is slowly added to a starch with a solids content of approximately 40% while stirring vigorously. The acid formed in the reaction is neutralized with soda ash while the reaction temperature is controlled either by cooling or by the rate of hypochlorite addition. The creation of organic chlorine compounds cannot be avoided during these reactions, and oxidized potato and maize starches contain, respectively, 20–100 ppm and 60–400 ppm organic halides (OX).

Starches treated with hypochlorite give pastes with lower viscosity and higher transparency than the corresponding native starches. Retrogradation tendency is reduced greatly by reduction in the carboxyl group content, with the result that the paste remains clear and liquid even after cooling.

Thermally modified starch. Thermally modified starches contain no OX; they are viscosity stable (Figure 4.7) and non-ionic. The stability of the starch paste is achieved because the macromolecules, which are hydrolysed at high temperature, undergo a transglucosidic rearrangement in the later part of the process. The amylose fragments are structured in such a way that amylopectin-like arrangements occur.

Starch ether. The most important group of chemically modified starches is the starch ethers. A wide spectrum of ionic and non-ionic starches can be produced by reaction with appropriately substituted epoxides, which are either used directly or are accessible *in situ* from the corresponding chlorohydrins (Table 4.2).

Hydroxyalkyl starch. Non-ionic hydroxyalkyl starches with degrees of substitution in the range 0.01–0.1 can be obtained in aqueous solution by reaction with ethylene oxide or propylene oxide.

The alkyl substituents considerably reduce the tendency for hydrogen bond formation and thus for retrogradation, with the result that the viscosity of the pastes is exceptionally stable in the event of cooling or long periods of storage. At the same time, the pastes possess a high water retention capacity and good binding power. These qualities, together with

Table 4.2 Methods of starch ether production

Structure:

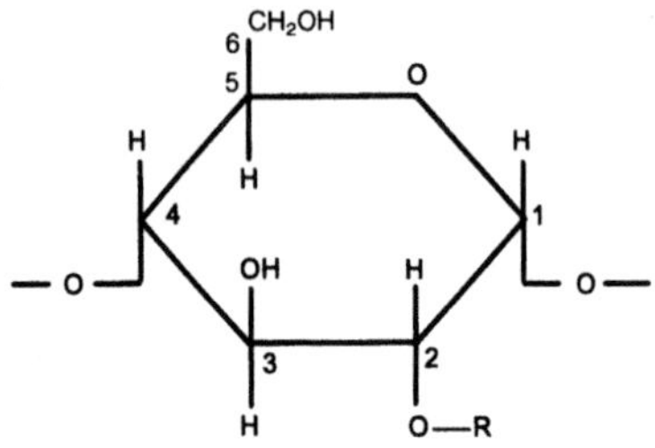

Reagent:	*Functional group, Ri:*
3-Chloro-2-hydroxy-propyltrimethyl-ammonium chloride (CHPT)	Cationic quatenary ammonium chloride
Sodium monochloroacetate	Anionic carboxylmethyl starch
Propylsultone	Anionic sulphonate starch
Calcium maleate	Anionic dicarboxylethyl starch

Structure:

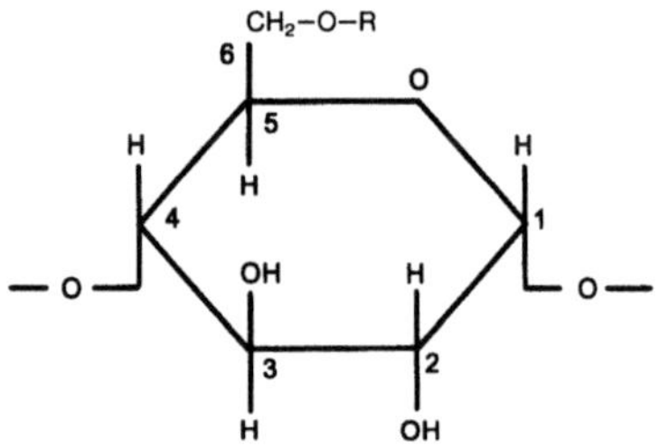

Reagent:	*Functional group, Ri:*
Ethylene oxide	Non-ionic hydroxyethyl starch
Propylene oxide	Hydroxypropyl starch
Epoxy propanol	Dihydroxypropyl starch

the considerable improvement in film-forming characteristics compared with native or oxidized starches, make this type of starch derivative particularly suitable for surface applications requiring these characteristics coupled with barrier properties to organic solvents.

Ionic starch. There are three categories of ionic starches: cationic, anionic, and bi-ionic or amphoteric, containing both cationic and anionic groups. The most important of these is cationic.

In order to incorporate cationic groups in the starch molecule, it is necessary to activate the starch chemically to the cationization reagent. This is done in the aqueous starch slurry by adjusting the pH to 11.5–12.5 with alkali and adding a cationization reagent such as 3-chloro-2 (hydroxypropyl)-trimethyl ammonium chloride (CHPT).

The higher the reaction temperature chosen, the faster the reaction process between starch and cationization reagent. At the same time, however, excessively high temperatures lead to side reactions and undesirable hydrolysis products as well as to inhomogeneous cationization of the starch granule. For this reason, a reaction temperature of 40°C–50°C is maintained. In addition, it is necessary to prevent the starch granules from partially dissolving at this temperature and pH. For this reason, salts are often added to inhibit gelatinization.

Owing to their cationic charge, the starches obtained possess high substantivity to anionically charged fibres, fillers or pigments.

Starch ester. Acetylation of granular starch in aqueous alkaline suspension is performed with acetic anhydride as the acetylation reagent. Esterification (Table 4.3) can be combined with other modification methods such as cross-linking, acid degradation or oxidation in order to obtain products tailored to the specific application. They are significantly less stable with respect to chemical influences than are the extremely stable starch ethers. For example, at pH 11, an acetylated starch with an acetyl content of 1.8% is completely saponified within four hours.

Excellent visco-stability, easy gelatinization, and thus problem-free handling during use and storage, as well as good film properties, are characteristic of starch esters and ethers.

Table 4.3 Methods of starch ester production

Structure:

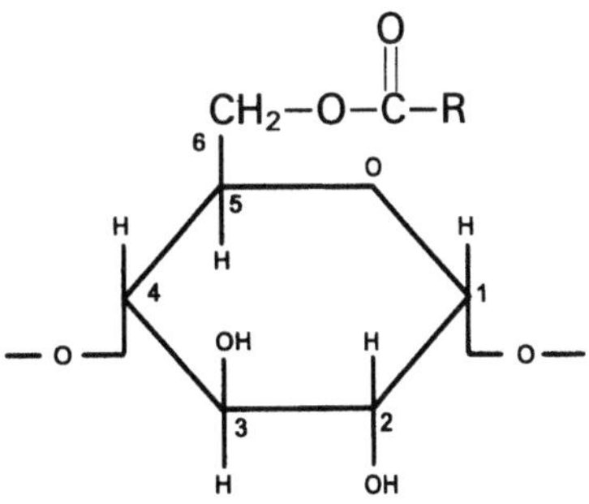

Reagent:	*Derivative, R–C=O:*
Acetic anhydride	Acetyl starch
Inorganic phosphates	Monophosphate starch
Carbon disulphide	Xanthane starch
Urea	Carbamate starch
Sulphur trioxide–amine complex	Sulphate starch

The viscosities of cationic starches, starch ethers and starch esters can be reduced to the desired level in the same ways as for regular starches.

4.4 Surface application methods

There are several methods for applying starch paste or starch-based coatings. In this chapter only those of importance for starch will be described and even then only insofar as these parameters influence the quality of application.

4.4.1 Size press

The most common surface application method remains the size press, consisting of two rolls pressed together to form a nip, where a pond of the liquid to be applied to the paper surfaces is formed (Figure 4.10). The paper passes through this pond. Excess starch is squeezed off between the two rolls, which may have different diameters and be covered with materials of different hardness and surface tension.

The liquid in the size-press nip is always in movement as the layers in contact with both the paper and the size-press rolls move at the same speed as each of these (Figure 4.11). This inevitably gives rise to turbulence.

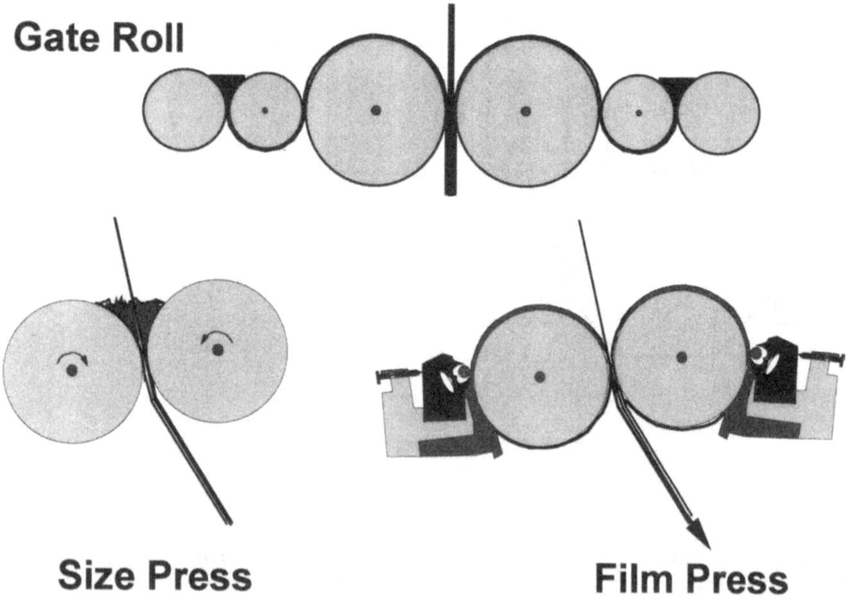

Figure 4.10 Surface application methods.

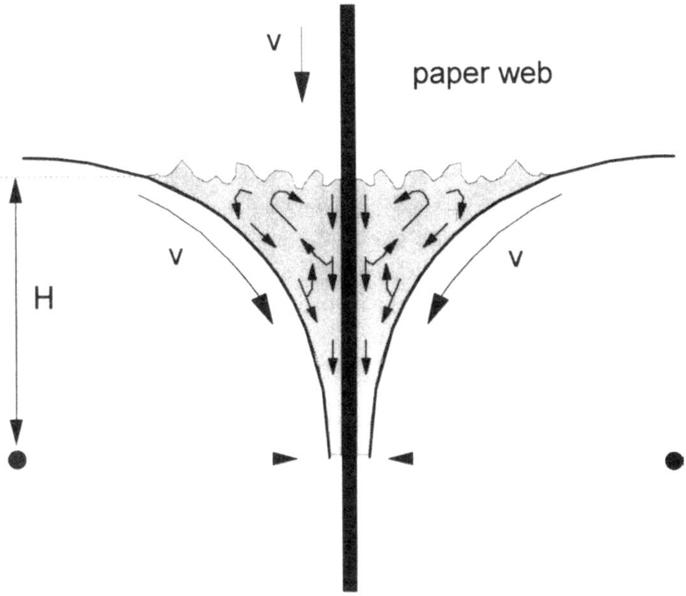

Figure 4.11 Hydrodynamic forces in the size-press nip. H = roll radius; v = speed.

After entering the nip, the liquid can either be pressed into the paper or be rejected in the opposite direction to the paper travel. The turbulence in the size-press pond will depend on the pond depth and width, the speed of the paper machine, and on the rheology of the size-press liquid.

Now they understand these factors, it has become normal for manufacturers to build size presses with larger and larger diameter rolls as papermachine speeds increase. Size-press speeds up to 1200 m min^{-1} can be achieved today using a first, and normally softer, roll diameter of 1.6 m. The other roll would in such a case be smaller: of approximately 1 m diameter. The hardnesses of the rolls would be approximately 20 P&J units and 0 P&J units, respectively, and the rolls would be mounted horizontally in order to build the greatest possible pond depth.

The hardness of the rolls and the pressure between them determine the nip width, which influences the penetration of the size-press liquid into the paper. The size press rolls have to be pressed together firmly enough to overcome the hydraulic pressure which is built up by the size-press liquid in the pond.

In spite of its widespread use, the size press nevertheless suffers from several important disadvantages:

- high energy requirement for afterdrying owing to the high fluid absorption;

- turbulence in the nip pond at high speeds which may cause non-uniform coating;
- the need to set the viscosity of the starch solution according to the hydrodynamic forces in the nip;
- fibre swelling resulting from excessive water absorption into the sheet, causing increased roughness;
- the running properties in the size press necessitate a certain minimum wet strength of the paper;
- the fluid absorption in the size press varies as a result of the moisture content of the incoming paper; in order to ensure a uniform moisture cross-profile, drying to a moisture content of approximately 1%–3% is necessary for graphic papers and to approximately 7%–12% for packaging papers.

Nevertheless, speeds of over 1000 m min^{-1} are achieved today using larger roll diameters (as described above), and using low-viscosity starches at low concentrations (3%–5% solids content). However, it seems provisionally that the conventional size press has probably reached the end of its development.

4.4.2 Apron size press

One approach to modifying the size press has been an attempt to reduce the turbulence in the size press pond by separating it from the high-speed surfaces of the rolls and paper by baffles of plastic material. This design became known as the 'apron size press', but it was demanding in terms of maintenance and cleaning (Figure 4.12).

4.4.3 Foam-coat size press

Other trials were aimed at foaming the starch solution in order to permit higher starch concentrations and thus lower water quantities to be coated onto the paper. Such applications became known as foam-coat size presses and allowed speeds of up to 1160 m min^{-1} to be achieved on test machines. However, the surface coating was uneven. In practice, only applications at low speeds (approximately 300 m min^{-1}) are known.

4.4.4 Gate-roll coater

The use of coating equipment for direct or indirect starch application permits metering of the supplied starch, while avoiding some of the turbulence. Premetering via roll systems – for example, the gate-roll coater (Figure 4.10) – has the advantage that uniform and reproducible premetering is possible only if the size suspensions have a relatively high

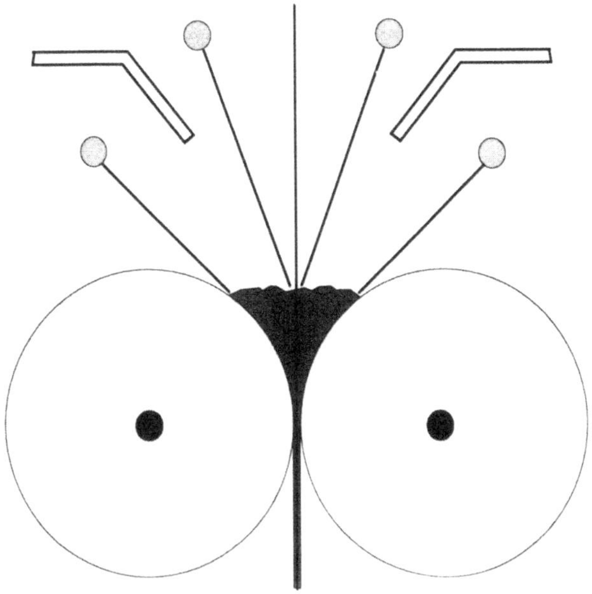

Figure 4.12 Apron size press.

viscosity. This is because of the hydrodynamic operating principle of the metering system.

In the case of a conventional size solution, this means solids contents of up to 20% must be used. However, a size concentration of 10% or less is desirable in order to obtain penetration into the sheet. Solids contents of 4%–5% can be used if non-degraded starches of excellent viscosity stability, i.e., starch ether or ester, are used.

Other disadvantages of this technology are as follows:

- the application weight depends entirely on premetering;
- the equipment is technically complex owing to the large number of rolls;
- the premetering rolls must be cambered and can be used only at a specific operating point;
- it is difficult to avoid deposits on the rolls.

4.4.5 Metered size press

In order to overcome the turbulence in the size press pond and the above-mentioned problems, the blade metered size press was introduced in 1982. The further development of this improved application system led to what today is called the metered size press (SpeedSizer, SymSizer, Film Press (Figure 4.10) TWIN-HSM, GRIC or gate-roll inverted coater).

The size press rolls now are used as applicator rolls (as in the gate-roll system). A film is applied to one or both of these rolls by means of a jet or fountain applicator system. The film is metered by either a profiled rod, or a smooth rod in the case of coating colour application. Blade metering, which was how the technology first developed, is also still used.

The applicator rolls in a metered size press are much softer than those of a size press. The first rolls used for starch paste application had hardnesses of 15 P&J and 25 P&J, and they both had the same diameter, approximately 1250 mm. The hardnesses of the applicator rolls are lower for coating applications.

In the size press the paper could be run only in the direction towards the floor. However, in the metered size press the paper direction can be either up or down depending on the best conditions for the different machines. Additionally, a new configuration of metered size press has been introduced to the market in which only one side of the paper is treated at a time, in separate units, like normal blade-coating stations.

The metering element may consist of a blade, rod or grooved rod. If premetering is performed with doctor blades, a film is applied to the roll with a premetering device similar to a short-dwell blade coater instead of by a roll premetering system (Figure 4.13). The film is transferred to the paper in the nip without pond formation and pressed into the paper; though the hydrodynamic pressure is low. Such metered size presses operate at speeds up to 1300 m min^{-1}, and pilot systems have already reached 2200 m min^{-1}.

The main advantage of metered size-press installations is the possibility of setting a defined application quantity at high papermachine speeds. In addition, the quantity of water which has to be evaporated in the after-drying assembly is considerably reduced (Figure 4.14). This leads to more favourable energy costs and improvements in running efficiency owing to the reduced tear sensitivity of the paper.

The metered size press is a very flexible piece of equipment. If, on the one hand, the emphasis of starch coating is on strengthening the paper surface, this demand can be satisfied by the film press without having to reduce the papermachine speed. If, on the other hand, the objective is more penetration through the whole thickness of the paper sheet, operation of a size press or metered size press in pond mode is possible for high-grammage papers. The use of the pond mode means that more starch is supplied by the premetering system than is absorbed by the paper, thus causing collection of the starch excess in the nip.

4.4.6 TWIN-HSM

Experience with the use of soft applicator rolls in the liquid application system (Section 4.4.7) led to the introduction of the TWIN-HSM coater.

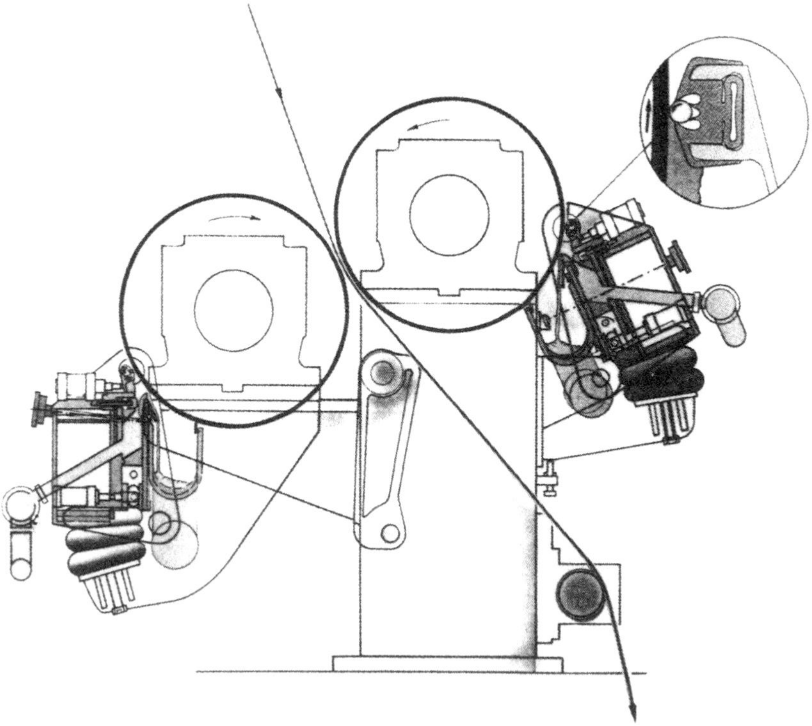

Figure 4.13 Film press: Jagenberg.

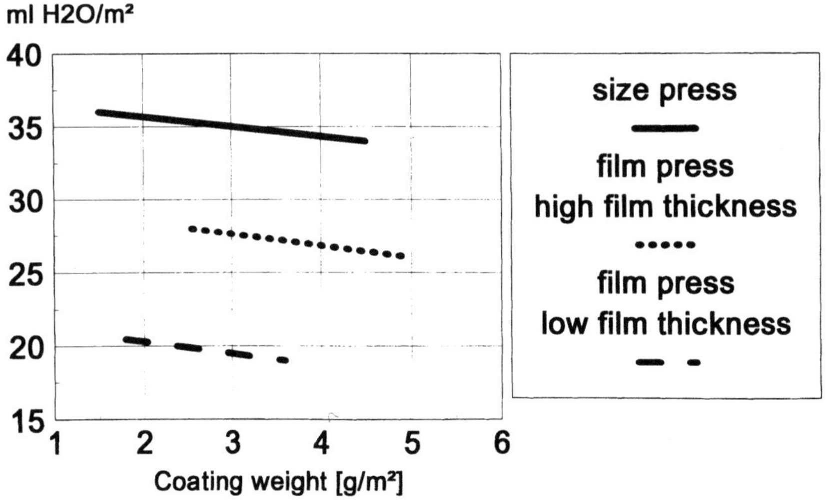

Figure 4.14 Water quantity to be evaporated as a function of coating weight.

As rubber hardnesses of 90–110 P&J are used on the applicator rolls and the wire-wound metering rolls are of large diameter (400–700 mm), the starch or coating colour is applied through a combination of volumetric and hydrodynamic metering. This leads to a long service life for the rolls.

The web normally runs upwards to facilitate cleaning, allowing the circulating rate of the applied suspension to be reduced to around five times the rate of application to the web.

On each side, the coating is fed into the pond between the HSM roll and the transfer roll from a distributor pipe fitted with holes across the full machine width. The pipe itself is double-walled to allow for cooling, which encourages condensation or humidity to form on its surface. This helps keep the pipe clean during operation.

The unit operates with a very low specific nip pressure, which, coupled with the soft rolls used, makes it possible to meter the starch or coating on the surface of the paper. If greater levels of starch penetration are required (for example, to improve internal bond) harder rolls and higher nip pressures can be used.

The TWIN-HSM is in use today at speeds up to 750 m min^{-1} on recycled base at weights as low as 35 g m^{-2}. Speeds of 1400 m min^{-1} have been achieved on pilot installations coating wood-containing grades.

4.4.7 *Liquid application system*

Another application unit, more usually used to add water to the back side of one-sided coated paper to prevent curl, is the liquid application system

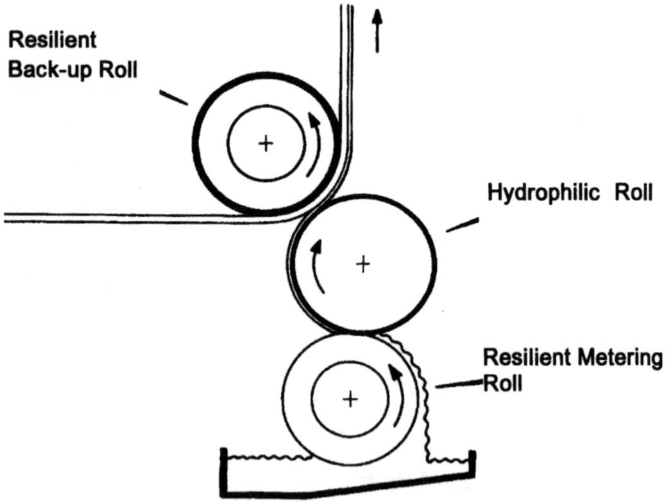

Figure 4.15 The liquid application system.

(LAS) of BTG, Säffle, Sweden. The LAS coater can be used to add starch solutions of low viscosity, too. The LAS coater (Figure 4.15) uses a hydrophilic transfer roll, which is chromium plated and treated by a proprietary etching process. The treated roll can carry wet films 3–15 microns thick, without slinging or spattering, at speeds up to 1700 m min^{-1}. Liquid is applied to the hydrophilic transfer roll by a resilient rubber-covered metering roll running in a supply pan. Varying the speed of the metering roll maintains a flooded nip to prevent air intrusion into the starch solution. Pressure in the metering/transfer roll nip is controlled to provide a uniform, unbroken film on the transfer roll. The hydrophilic transfer roll is run in the same direction as the web to apply the liquid film to the web, though with a speed differential between them to minimize film-split pattern on the sized sheet. Pick-up is controlled by varying transfer roll speed and the pressure between metering and transfer rolls. A resilient rubber-covered backing roll is used to ensure uniform contact between the sheet and the hydrophilic transfer roll.

4.5 Starch properties required for size-press application

The most important aim of starch application is improvement of the mechanical strength both at the surface and within the body of the sheet. This can be achieved by penetration of the sheet structure by the starch solution. The effect of the starch can be measured by means of one or more of the following parameters:

- tensile strength;
- plybond strength;
- pick resistance (Dennison wax, IGT);
- ring crush test;
- flat crush test;
- stiffness;
- printability;
- runnability.

Surface application of starch thus primarily serves to improve certain paper properties and permits the economy of paper manufacture to be improved by optimizing the cost of the raw material used.

In order to improve the desired property it is necessary to understand the behaviour of starch during the size press treatment and how different starch modification influences both the runnability of the size press and the paper properties.

Starch data for two papermachines equipped with size presses are shown in Table 4.4. Both machines produce wood-free uncoated paper. It can be seen that the liquid volume absorbed differs between the heaviest and the

Table 4.4 Typical size-press conditions in modern papermachines

	Area weight (g m^{-2})			
	45	60	80	100
Papermachine A				
Size-press speed (m min^{-1})	> 900	800	650	
Starch concentration (°Bx)	4	6	7	
Starch viscosity at 50°C, 100 rpm (mPa s)	20	28	30	
Starch pick-up				
solution (ml m^{-2})	50	55	57	
dry (g m^{-2})	2	3.3	4	
Papermachine B				
Size-press speed (m min^{-1})		> 900	800	600
Starch concentration (°Bx)		7	9	12
Starch viscosity at 50°C, 100 rpm (mPa s)		30	40	50
Starch pick-up				
solution (ml m^{-2})		43	44	50
dry (g m^{-2})		3	4	6

lightest quality by about 14%. The reason for the different applied quantity of liquid is probably the lower absorption capacity of the lighter papers. This will be discussed in more detail in the following.

4.5.1 Influence of viscosity

The influence of starch viscosity together with subsequent absorption is shown in Figure 4.16. The forces which counteract fluid transport grow tremendously as the viscosity of the penetrating fluid increases. In this case the viscosities are set on the basis of the desired final dry pick-up. However, it should be of no significance for the absorption rate whether the viscosity has been set on the basis of the dry content or the chain length. The amount of starch stock absorbed is determined solely by its viscosity.

However, there is some question as to how best to measure the viscosity of starch paste. Brookfield viscosity, the normal measurement used in the paper industry, does describe the size-press behaviour of a known starch paste in an appropriate way. The viscosity of the starch solution should be measured at as high a concentration as possible, as low concentrations do not differentiate between starches well enough. The measurement is sensitive enough to control the size-press make-up system; however, it does not provide enough information to predict the effect of replacing one starch by another.

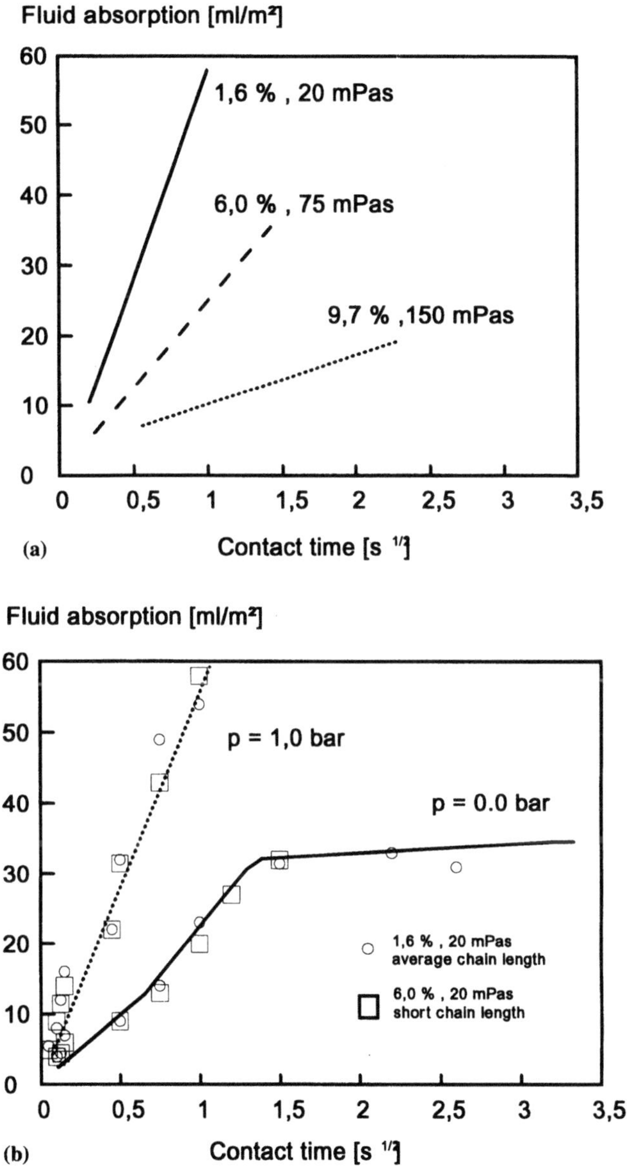

Figure 4.16 The influence of viscosity on starch absorption.

It has been reported from a mid-European mill producing wood-free paper that the size-press pick-up decreased by 5%–10% (depending on the paper quality) when a low viscosity oxidized potato starch was replaced by a maize starch of the 'same' Brookfield viscosity. The solids contents of the starch pastes were between 6% and 8%; the viscosity was approximately 25 mPa s at 100 rpm and 50°C; the papermachine speed was 500–700 m min^{-1}.

Extensive trials over several three-month periods in a Scandinavian mill producing the same paper qualities showed almost the same results with thermochemically converted maize and potato starch (see Table 4.5).

It seems to be evident that the molecular weight distribution of the starch plays a dominant role in this behaviour. Figure 4.9 shows that the chain-length distribution of starches of similar viscosities can vary considerably. It seems to be logical that the lower molecular portions of the starch are more easily absorbed and that these portions also penetrate faster. This could be the explanation for the differences between research investigations and practical results from fast-running papermachines, with regard to the relation of starch pick-up and viscosity level.

The above-described findings do not mean that maize starch is better than potato starch or the opposite. They show that starch solutions of the same Brookfield viscosity behave differently and that other measuring techniques have to be used if one has to rely on laboratory investigations only.

Solutions of maize starch have a shorter structure than those of potato starch. This structure difference can be seen by pouring the starch solution from a beaker very slowly. A very fine filament can be formed with potato starch, whereas the maize breaks up into droplets. A Bohlin rheometer can also detect the difference. The Bohlin rheometer can measure the starch viscosity at low shear rates and is a good tool for investigating starch paste. The shorter structure of the maize starch solutions leads in some size presses to a lower splashing tendency and to less misting during film-press coating.

When the viscosity is increased from, say, 40 mPa s to 80 mPa s, measured at the same solids content and temperature, the liquid pick-up in

Table 4.5 Comparison of maize and potato starch in size-press application

Starch	Concentration (%)	Starch pick-up (%)	Sizing solution consumed per ton of paper (l)
Potato	6	3	500
Maize	6	2.7	450
Maize	7	3	430

- 50 l less water to evaporate;
- 3 kg less starch consumed with maize.
- 70 l less water to evaporate with maize.

Table 4.6 The relation between starch viscosity and liquid add-on (speed = 500 m min^{-1}; diameter of size-press roll = 60 cm)

Dry solids content (%)	8	12	8	12
Viscosity (mPa s)	20[a]	30[a]	43[b]	80[b]
Liquid add-on (ml m^{-2})	49	49	43	41

[a]Low viscosity.
[b]Medium viscosity.

the size press may increase because of the increased hydraulic load. At speeds below 600 m min^{-1} and with small diameter size-press rolls, however, the applied amount of liquid will decrease as the viscosity increases (Table 4.6).

4.5.2 *Influence of solids content*

The starch pick-up increases almost linearly with the solids of the size-press solution on fast-running papermachines. This means that less water is added to the paper. However, less water to evaporate does not necessarily mean that steam consumption in the after-drying section can be decreased. The opposite is the case. More starch pick-up means that the applied water will be more difficult to evaporate. Therefore the steam consumption will rise or the machine speed will have to be reduced (Figure 4.17). The water-holding capacity of the starch is greater than that of the fibres.

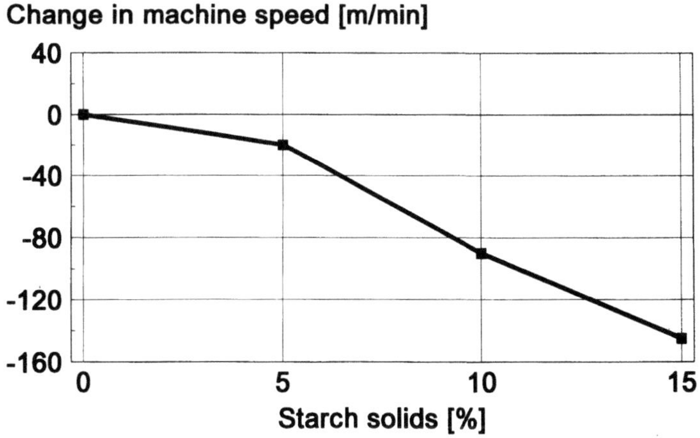

Figure 4.17 The influence of size-press solids.

4.5.3 Penetration

The penetration of starch depends on the following factors:

- viscosity of the starch;
- solids of the starch;
- paper structure;
- paper moisture;
- fibre swelling behaviour;
- linear pressure of the size-press rolls;
- hydrodynamic pressure of the size-press liquid;
- molecular weight distribution.

These factors have not been investigated sufficiently thoroughly that the desired answers can be given. However, industrial experience gives enough information to understand the first four.

The size-press penetration and pick-up decrease when the moisture of the paper entering the size press is reduced from 10% towards 1%. A wood-free paper will probably be at 1%–3% moisture when entering the size press. At higher moisture content it is difficult to achieve an even enough moisture profile across the web. An uneven profile means quite different starch pick-up, resulting in unacceptable profiles at the end of the machine.

Paper for corrugating boxes will enter the size press at approximately 10% moisture. The reason for this is that this type of paper needs a high starch pick-up and good penetration in order to develop the desired strength properties (Concora test and ring crush test). The moisture profile at the end of the machine is indeed difficult to control, though an acceptable quality for the box manufacturer is achieved (more details in Section 4.9 on packaging paper).

The penetration depth of starch can be determined by the microtome sectioning technique or by measuring different paper strength characteristics of the produced paper.

A low-viscosity starch penetrates paper more than does a starch of higher viscosity. This can be seen in the microtome section shown in Figure 4.18, and in the resulting surface strength figures illustrated in Figure 4.19.

A wood-free paper of 100 g m^{-2} can be completely penetrated, independently of the sizing of the paper, provided the porosity of the paper is high. At greater paper weights it can be anticipated that the starch will not penetrate the paper completely. The internal bond of these papers might have to be improved by addition of 0.8%–1.5% cationic wet-end starch, added to the thick stock.

The solids content of the starch influences the penetration behaviour only if the starch viscosity is drastically increased by increased solids. This can be seen if the penetration of an acid-modified starch which has high

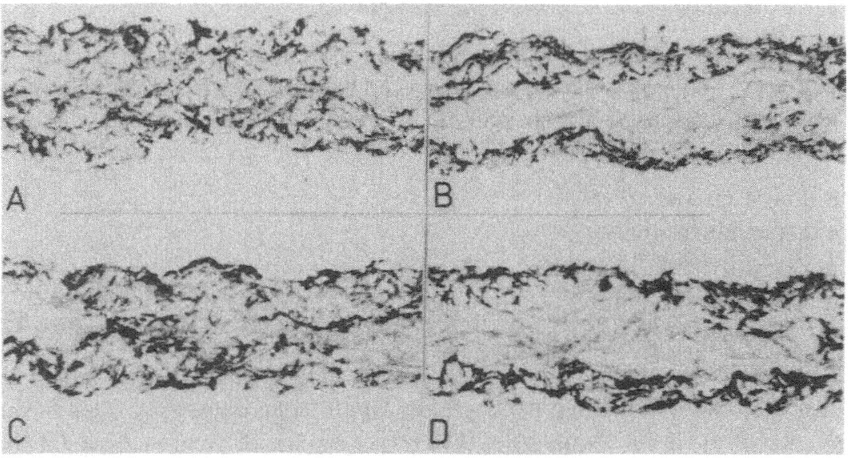

Figure 4.18 Microtome sections of 80 g m^{-2} wood-containing papers. A, B = size-press treated: C, D = film-press treated.

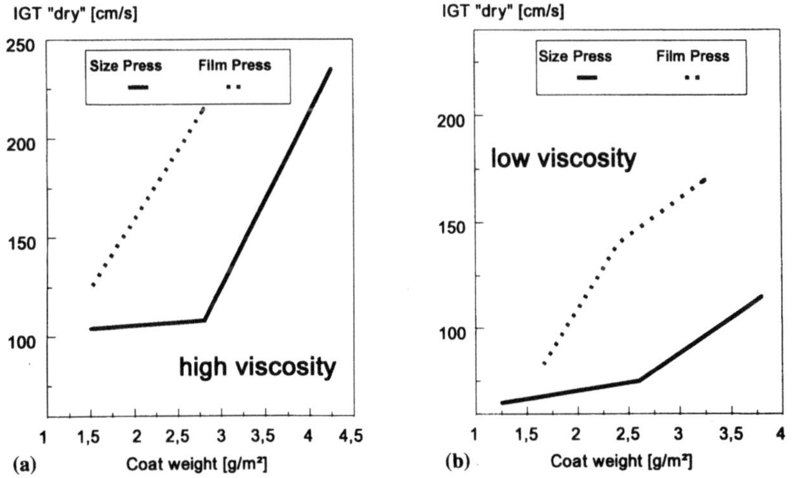

Figure 4.19 Correlation of starch viscosity and surface strength in 80 g m^{-2} wood-containing paper.

retrogradation tendency (set-back) is compared with that of an oxidized starch with lower set-back. At low solids (that is, 5%) there will be no difference; however, at 10% solids the acid-modified starch will stay more on the surface of the paper.

It is obvious that a paper where the fibre composition is mainly ground wood will swell rapidly when it is wetted in the size press; and the number and the size of the pores is reduced. These papers are more difficult to penetrate with a starch solution.

4.5.4 *Influence of papermachine parameters*

The size-press pick-up increases as the speed of the papermachine is increased. This is caused by the increased hydraulic forces in the size-press nip. The increased starch pick-up at higher speeds is, however, also dependent on:

- starch viscosity;
- paper absorptivity;
- size-press roll diameter;
- size-press nip load.

Also, it is only possible to apply general rules.

As the hydraulic pressure increases with increased viscosity it should be expected that the pick-up will increase with increased viscosity. This is true for size presses with large rolls. The effect is even more pronounced if the nip load of the size-press rolls is not too high at the same time. Papers of high weight (more than 100 g m^{-2}), with an open structure and made predominantly from fibres which do not swell quickly show a higher coat weight increase as speed increases compared with closed papers with a limited absorption capacity.

Smaller diameter size-press rolls and harder size-press roll surfaces have higher linear pressures in the nip. These presses do not show so much influence of speed increase as do those with bigger rolls or softer covers.

Little evidence is available regarding the influence of linear pressure of the size-press rolls on the penetration of starch solution. It is said by some papermakers that they achieve a better penetration of the starch if they use small steel rolls where a high linear load can be obtained. However, it has been reported that the penetration does not change if the nip load is increased from 25 kN cm^{-1} to 45 kN cm^{-1}.

Paper producers will seldom see any influence of nip pressure on starch pick-up or penetration over the speed range in which they normally operate. Certainly, it is necessary to avoid allowing the hydraulic pressure to exceed the nip pressure. If this takes place an immediate break will result.

4.6 Starch properties required for metered size-press application

As described above, there are in principle two main parameters which can be controlled during size-press application:

- the solids content of the starch paste;
- the viscosity of the starch paste.

Metered size-press application adds another parameter to these two:

- the film thickness on the applicator rolls.

However, the main difference between size-press and metered size-press application of starch is the absence of the pond in the applicator roll/paper nip. That means that the hydraulic forces of the size-press nip are not present; or, if a 'minipond' is present, they will be very low. The result of this is that the penetration of the applied liquid will be reduced compared with the size press. At the same time, the absence of a pond makes it possible to work with much higher viscosities on the metered size press.

The maximum possible film thickness is dependent on the absorptivity of the paper used; more absorbent papers can accommodate a thicker film on the applicator rolls before a pond, which causes splashing, is created.

Independent of the paper quality, savings in drying energy can be achieved when switching from size-press to metered size-press application. However, it is true both for size-press and for metered size-press application that the higher the concentration, the more starch is applied and hence the more difficult the sheet is to dry.

As shown in Figure 4.20, 3 g m^{-2} of starch can be applied with either an 8% size-press solution or with a 10%–16% metered size-press solution, depending on the profiling of the metering rod used. In this way, the amount of water to be evaporated can be reduced by 22%–54%. It is therefore clear that the metered size press should be installed closer to the reel-up.

It is also found that more stable moisture profiles are obtained with the metered size press, although the paper moisture level at the entry to the metered size press can be higher (typically, 1%–2% for the size press and 5%–7% for the metered size press).

Differences in the dependence of applied weight on the solids content are difficult to detect when using two starches with completely different

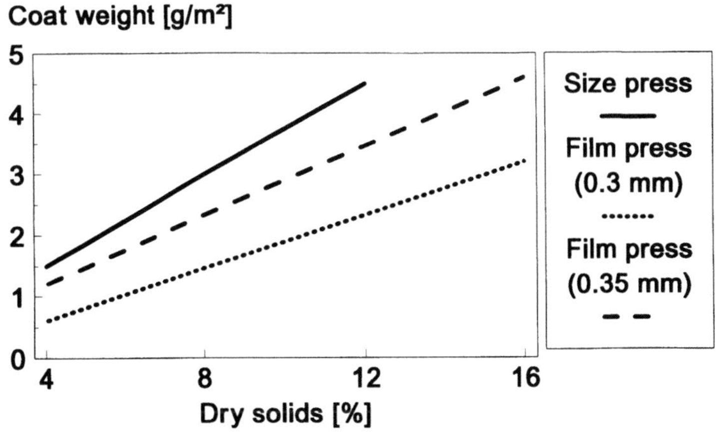

Figure 4.20 Starch application on 80 g m^{-2} wood-containing paper.

viscosities. There are, however, slight differences, as the film-splitting effect when the paper leaves the applicator rolls is different. A film of a low-viscosity starch applied at 10% will be split so that approximately 90% is applied to the paper and 10% returns to the metering rod. Higher starch viscosities can lead to a 70%:30% split in the divergent zone. A relationship between starch viscosity and the amount of liquid starch applied to the paper as a percentage of the amount of liquid applied to the applicator rolls (known as the transfer ratio) is shown in Figure 4.21.

The penetration behaviours of starches of different viscosities are different too. Starches of lower viscosities penetrate more. This is shown clearly in Figure 4.18 where results from applying starches of different viscosities to an 80 g m^{-2} wood-containing paper in a size press as well as a metered size press are shown. The high-viscosity starch (24% solids at 50°C and with a viscosity of 1500 mPa s) penetrates only partially both in the metered size-press and size-press application, which are essentially the same. On the other hand, the starch of much lower viscosity (110 mPa s, measured at the same concentration and temperature) penetrates the paper completely when applied at a size press. Penetration after metered size-press application is good too, however, because of the rapid swelling of the groundwood fibres. It is only a little less than for the size-press treated paper.

The physical testing of the papers described above produced the expected results. Slightly higher tear strength values are obtained with size-press treatment than with metered size-press treatment. However, the results differ with respect to surface strength (Figure 4.19). Here, the metered size press proves to be superior. The increased starch concentration in the layers close to the surface improves the surface strength for both the low-viscosity and the high-viscosity starch solution.

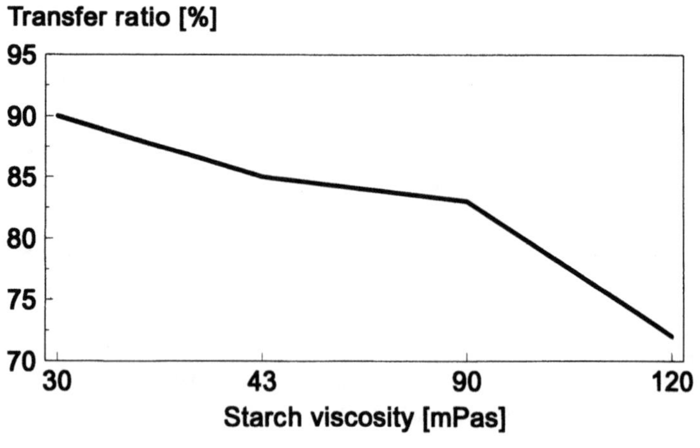

Figure 4.21 Relation of starch viscosity and film splitting during film-press application.

Recycled fibre papers for packaging are size-press treated with starch in order to improve the Concora medium test (CMT) value for fluting papers and to improve the burst values of liner qualities. To improve the CMT, it is believed to be of advantage to fill the paper with starch. This means that low-viscosity starch products should be used when these paper qualities are produced with metered size-press treatment. Even so, it is difficult to achieve the same improvements from starch application with metered size-press treatment. It is necessary to apply more starch with the metered size press to compensate for the reduced penetration of the starch.

It is reported that trials with 16%–24% solids in a metered size-press solution leading to starch pick-up of approximately 9 g m^{-2} give the same strength as would be expected from a size press. It can be said that good CMT values can be achieved without complete penetration of the starch. Penetration of greater than 30% from each side is considered to be sufficient.

The thickest applied film on the applicator rolls gives the best results, since this allows the required pick-up to be achieved by a thick film of low concentration. This in turn allows for the best penetration.

Burst improvement of packaging paper can more easily be achieved by metered size-press treatment. Starches of higher viscosity will give better results than will low-viscosity starch.

Comparing low-viscosity with high-viscosity starch, low-viscosity starch allows the same ring crush test (RCT) levels to be achieved as the high-viscosity starch, but at a higher pick-up. Low-viscosity starch allows very high RCT levels to be achieved, at correspondingly high pick-ups.

Higher speeds are always possible with a metered size press than with a size press while achieving comparable paper strength; however, more starch has to be applied. Low-viscosity starches allow higher machine speeds than do the high-viscosity types.

If the conditions are compared under which the greatest improvements of CMT, burst and RCT are achieved it is obvious that a high-viscosity starch on the size press is the most economic with regard to starch effectiveness.

The balance between higher starch costs and higher productivity needs to be established before any decision to incorporate a metered size press is taken.

4.7 Starch properties required for blade application and as a coating colour binder

4.7.1 Blade application of starch solutions

Starch application with on-machine short-dwell coater units (short-dwell-time applicator (SDTA)) is not often carried out, though it can certainly be done.

Starch application weight is influenced only by the solids content of the starch solution and not by the starch viscosity or blade pressure of the SDTA. The solids content used in the industry for this type of application will vary between 3% and 10%.

Microtome sectioning of newsprint papers which had been surface treated with starch solutions of different solids and with two starches of quite different viscosity levels showed that penetration by the starch solution is only possible at low solids content. In Figure 4.22 it can be seen that a starch solution of high solids remains concentrated on the paper surface. Applied at, for example, 5% solids, the starch viscosity does not influence the penetration. At higher solids, e.g., 10%, the high-viscosity starch penetrates less than does the low-viscosity starch.

These findings demonstrate well how blade application of starch influences the technological properties of a paper (Figure 4.23):

- tensile strength increased with increased amount of applied starch;
- a 35% increase of stiffness was achieved with a pick-up of 0.5 g m^{-2} per side solids.

Increased tensile strength does not invariably lead to decreased Brecht–Imset tear values. Addition of high-viscosity starch increased the tear resistance of the newsprint paper, which was used as base stock whereas low-viscosity solutions did not. It is evident that the shear stress behaviour of fibre-to-fibre bonding in the paper is less influenced the more the starch

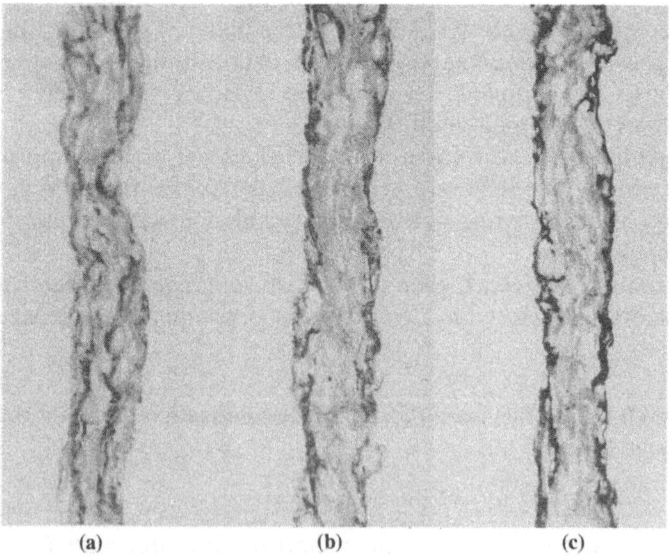

Figure 4.22 Microtome sections of papers treated with starch by means of a short-dwell coater unit. Starch solids contents: (a) 5%; (b) 10%, and (c) 15%.

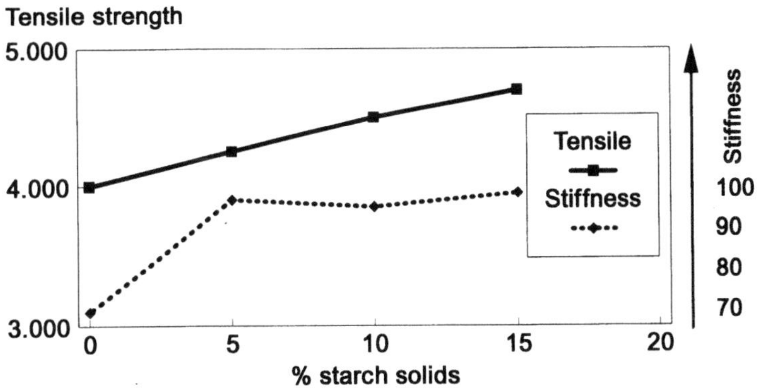

Figure 4.23 Influence of starch addition on tensile strength.

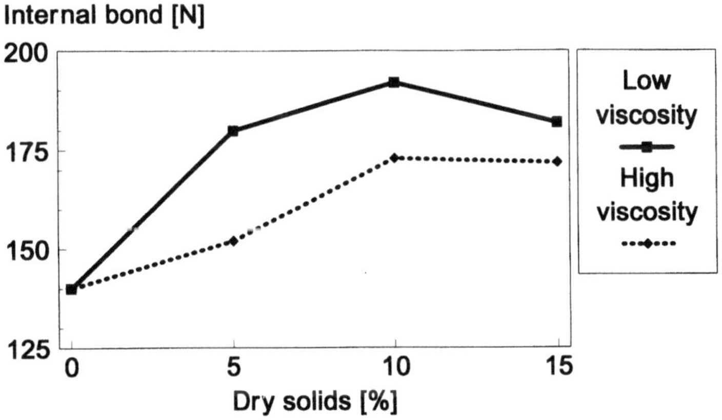

Figure 4.24 Influence of starch viscosity and solids on internal bond. -■- = low viscosity; —◆— = high viscosity.

is kept on the paper surface. These findings also correspond to the development of the internal bond values (Figure 4.24).

The surface strength increased at the same time with the increased starch addition (Figure 4.25). The tendency to linting and dusting will be reduced as the viscosity of the starch increases.

Offset printing of wood-containing paper shows that paper treated with high-viscosity starch will be printed without linting and dusting if 0.5–0.8 g m^{-2} of starch have been added to the paper surfaces. This can normally be achieved if starch solids of 3%–8% are used at the blade station. Figure 4.26 illustrates the surfaces of three newsprint papers with starch films of different weight applied to their surfaces.

The gloss of calendered paper does not appear to be influenced by the starch treatment; printing gloss, however, can be significantly improved.

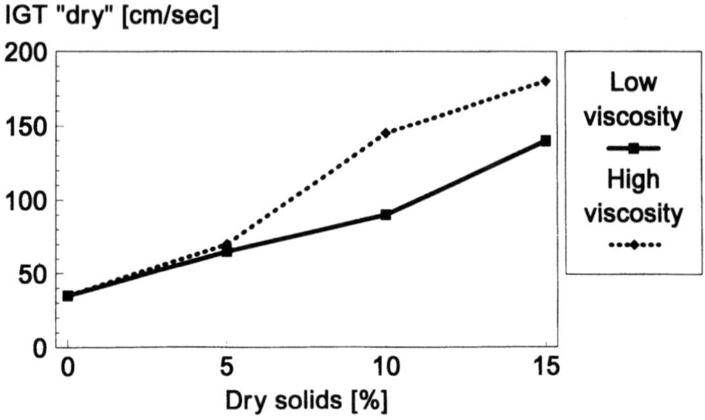

Figure 4.25 Influence of starch viscosity and solids on IGT 'dry'.

This will depend on how effectively a starch film can be formed on the paper's outer layer. It should be clear, however, that both opacity and brightness of paper will decrease when the fibre voids are filled by starch, though the brightness loss is reduced when high-viscosity starch is used (1.5 points compared with 2.5 points with low-viscosity starch).

In summary, then, surface treatment by blade with starch (or with carboxymethylcellulose (CMC) or other film formers, or with a lightly pigmented coating) can greatly improve surface strength and reduce linting and dusting. The loss of brightness and opacity must be compensated by the use of different pigments.

4.7.2 Starch as a coating binder

Starch is a very efficient chemical in coating colours both for binding pigment particles to each other and to the paper substrate, and for controlling rheology and water retention.

The amount of binder which is used in coatings depends to a large extent on the solids of the coating colour. In size-press coating colours, which can range from 45% down to 15% solids, the binder quantity can be in the range of 25–100 parts based on 100 parts of the pigment used. The relationship between solids and binder amount is shown in Figure 4.27. At lower solids contents, starch will be used as the only binder; whereas in higher solids coating colours, other binders may also be used, depending on the purpose of the applied coating. If this coating layer is a pre-coat it can be assumed that 70%–100% of the binder will be starch. The viscosity of a size-press coating will normally be around 100 mPa s at higher speeds, but must be increased to 400 mPa s at lower speeds. The running

(a)

(b)

(c)

Figure 4.26 Oblique views of the surfaces of three papers treated with increasing amounts of starch with a blade applicator: (a) base; (b) 0.4% g m^{-2} per side starch solution; and (c) 0.7% g m^{-2} per side starch solution.

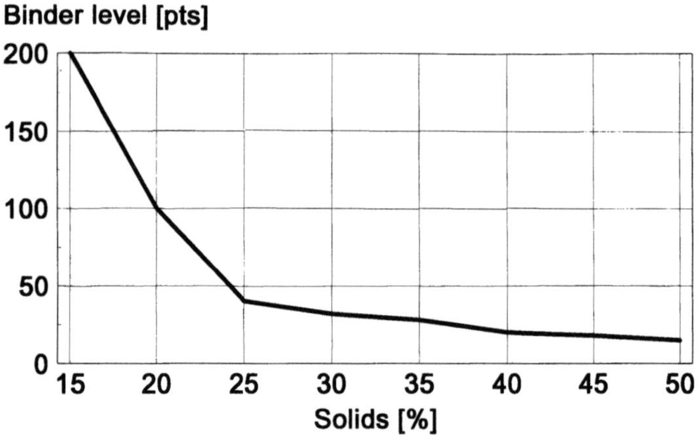

Figure 4.27 The binder proportion required in coating colours of increasing solids content.

conditions of the size press determine the ranges of solids content and viscosity of the starch that can be used.

A paper or board which is size-press coated, even at lower solids content, has improved surface strength and surfaces closed to such an extent that the next coating can be applied with a strong reduction of latex binder.

A lower starch:latex ratio is used if the paper is to be offset printed directly after the size-press coating. The ratio can be 50:50 or 30:70. The starch used for these applications will normally be the same as the starch used for pigment-free application in a size press.

Much wider ranges of solids content and viscosity are possible with metered size-press application of coating colours than with size-press application. Solid levels up to 65% can be applied. It is natural that the viscosity of the starch used in a formulation of 65% dry solids is much lower than for when the solids level is in the normal size-press range of 30%. The rheology of the metered size-press coating colour can be adjusted not only by means of the solids content of the coating and the quantity of starch but by the type of starch used. Stabilized cationic starches create a higher hydrodynamic pressure, desirable for the application of the coatings by smooth rods, than the enzymatic converted regular or oxidized starches. The starches which are used in coating colours for short-dwell blades differ from those used in blade coaters equipped with application rolls. The 4–6 parts of starch which are used for low-weight coat (LWC) offset colour together with 6–7 parts of latex need to create a high low-shear viscosity if the coating is applied by means of applicator rolls. Additionally, the water retention of the coating should not be too high. However, such coating colours need to have a low high-shear viscosity. All these parameters can be achieved only with difficulty, and it is necessary to work with starches which are specially designed for the different pigment systems. Starch esters of high viscosity can fulfil the needs.

For short-dwell coaters it is easier to select the starch, as the only demand on it is that the high-shear viscosity of the coating colour be low. Oxidized starches or thermally modified starches of low viscosity will fulfil this demand.

4.8 Starch properties required when applied by liquid application system

The LAS is normally used to apply water to the back side of one-side-coated papers to avoid curl. However, this coater is more versatile than that. As starch can be applied without a pond and under very low linear pressure, the starch solution will not penetrate and will be located at the paper surface. This technology should be used only where such characteristics are required. The quality of starch which can be applied depends on the starch solids. The viscosity level for the applied liquid solution is limited to a maximum of 200 mPa s. LAS technology can even be used to apply starch to upgrade a newsprint paper; an 11% solution of low-viscosity starch ether (of viscosity 30 mPa s) applied at 800 m min^{-1} gives a pick-up of approximately 0.5 g m^{-2}.

4.9 Size-press starch application to waste-based packaging paper

Starch addition of some kind is a prerequisite to producing fluting and liner grades based on recycled fibres with the required strength properties. Three principal technologies are available for this:

- wet-end addition;
- spraying;
- surface application.

The closed circuits of modern machines using recycled fibres are normally so loaded with anionic trash that addition of cationic wet-end starch does not result in the desired strength values.

Although the spraying technique is a very useful tool to improve paper strength, it is limited in how much starch can be added and needs quite high temperatures in the first drying cylinders. Additionally, air turbulence on high-speed wires (> 600 m min^{-1}) can disturb the regular distribution of the spray pattern.

For surface application, the size press is state of the art in the packaging paper industry. With a size press, only the parameters of starch concentration and intrinsic viscosity (on an average molecular weight basis) can be changed in order to achieve the desired physical strength in the most economical way.

The starch used for this purpose can be based on maize, potato or wheat; and the starch is normally viscosity-modified by means of enzymatic or thermochemical conversion in the paper mill. The level of solids for size-press application is in the range of 6%–8%. The viscosity levels are very different, however, depending on size-press configuration and roll diameter as well as papermachine speed. Levels of 25–30 mPa s, measured at 50°C, is one popular operating range; the other is to use a far higher viscosity, that is, 50–70 mPa s. Each mill needs to find the viscosity level which suits it best, considering both runnability in the size press and the absorption behaviour of the paper. If the Concora figures before size-press treatment of a 120 g m^{-2} paper are in the range of 120–140 N, applying starch will be expected to raise this by 15–20 N g^{-1} of added starch m^{-2}. A total of 4% of starch is normally added during size-press application to this type of paper.

The major drawback of the size press is that the machine speed has to be reduced by 15%–25% compared with an operation without a size press. The extent of this speed reduction depends, of course, on the parameters of the size-press solution applied. Thus:

- the higher the starch concentration (and thus the pick-up), the lower the machine speed;
- the higher the starch viscosity at a given concentration, the lower the machine speed.

The drying requirements are thus ruled generally by:

- the volume of sizing solution added to the paper in ml m^{-2} (more water to be evaporated);
- the water retention of the starch (the higher the viscosity the more difficult it is to dry);
- the degree of penetration of the sizing solution into the base sheet (the deeper the penetration the more difficult to dry).

A real decrease of drying requirement for surface sizing is only possible if techniques are available that limit the pick-up at increased solids levels (metered size press or one-sided size press).

4.9.1 *One-sided size-press operation*

One means of achieving a lower drying requirement for fluting papers is the one-sided size press. In this configuration, the inlet guiding roll of the size press is displaced, thus allowing a greater pond depth with splash-free running at higher concentrations and viscosities.

Comparing the results from a mill trial shows that the ideal situation is

Table 4.7 One-sided size-press operation on a 118 g m^{-2} fluting paper

Size-press operation	Starch data		Add-on (g m^{-2})	CMT 60 (N)	Effectivity (N m^2 g^{-2} starch)	Machine speed (m min^{-1})
	Dry solids content (%)	Viscosity (mPa s)				
Both sides	6.5	40	4.1	191	19	200
From felt side	12.5	112	5.2	181	13	250
From wire side	11.5	106	3.5	160	13	255

not completely achieved (Table 4.7). By changing from 6.5% dry solids in normal two-sided operation (giving 4.1 g m^{-2} pick-up), to one-sided application to the felt side with 12.5% dry solids in the size press, the pick-up can actually be increased.

Nevertheless, the machine speed can under such circumstances be increased by 15%–20% and the desired strength requirements can come close to those achieved with two-sided operation. Often it will be necessary to improve the strength of the paper by an additional introduction of 2% of wet-end or spray starch.

One-sided coating to the wire side, which is less 'open' to the sizing solution, would mean that the desired pick-up and therefore strength values would not be achieved.

At viscosities around 110 mPa s even at pond depths such as those achieved by increasing the size-press roll diameter or by one-sided size-press operation, the onset of splashing will limit the machine speed.

It should be pointed out that liner qualities are normally one-side treated in the size press with a blend of starch and sizing agent. The solid level will be approximately 8% and the sizing agent addition can vary between 10%–20%, based on dry starch. The outer paper side is treated. If both sides should be treated it can happen that the 'back' or inner side of the paper will be too closed and gluing problems can arise in the corrugating machine.

5 Rheology modifiers and pigment dispersants

J.B. CLARKE

5.1 Introduction

Coating colours are essentially colloidal mixtures of pigments, starch and latex particles which, as simple blends, would be totally unsuitable for coating onto base papers and boards. The smooth operation of the coating process, whatever its nature, often referred to as its runnability, is significantly dependent upon the careful control of the viscosity and rheological properties of the coating colour. This chapter considers the chemistry and behaviour of the two essential chemical additives used in combination with the latex, starch and pigments which, by careful selection, can achieve this control. The chemicals are: viscosifiers or rheology modifiers (also known as thickeners), and pigment dispersants.

The viscosity and rheology of a coating colour are influenced by many factors including:

- the total solids of the system;
- the binder particle size;
- the particle size distribution, shape and degree of dispersion of the pigment;
- the chemical type, molecular weight and dosage of the thickener.

In turn, the viscosity and rheology profoundly affect the following important parameters of the coating operation:

- the coating weight and uniformity;
- the pigment–binder distribution;
- binder migration;
- water retention behaviour;
- surface properties such as gloss and ink receptivity;
- the coating machine runnability.

5.2 General principles of viscosity and rheology

The study of viscosity and rheology of fluids is a complete science in itself and there are several sources of detailed information [1]; only the fundamental principles as they pertain to paper coatings will be considered here.

There are two important definitions: viscosity is defined as the resistance of a fluid to flow; while rheology is the study of flow behaviour under conditions of shear.

The effect of applying shear to a fluid by stirring or by forcing it through an orifice or other small gap between two surfaces is to induce a stress. The magnitude of this stress is related to the viscosity and to the rate of shear. The flow behaviour is usually represented graphically from results generated using a variety of measuring devices (Table 5.1).

Consider a model system in which a fluid is confined between two parallel plates separated by a distance z (as shown in Figure 5.1). One plate is fixed and the other, of area A, is moveable. If a force F, is applied to the

Table 5.1 Viscosity–rheology measuring devices

Rotational viscometers	Capillary
Brookfield	Ford/Din cup
Haake	Ubbelohde
Contraves	High-pressure capillary
Shirley Ferranti	
Rheomat	
Bohlin	
Carrimed	

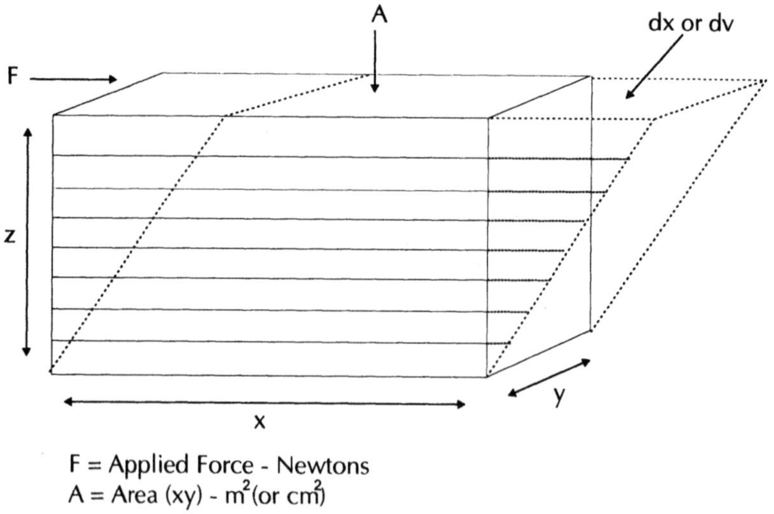

Figure 5.1 Model of a fluid between two parallel plates. z = distance between parallel plates; $A = xy$ = area of movable plate; F = force applied to movable plate; v = velocity of movable plate under force F; dv and dx = change in velocity and position, respectively, owing to application of force F.

moveable plate it moves sideways with a velocity v, taking the adjacent layer of fluid with it. However, the velocity of the fluid adjacent to the fixed plate is (nearly) zero, indicating that a velocity gradient has been introduced. This velocity gradient is referred to as the shear rate, γ.

It can be shown that, in a simple flow such as this, referred to as Newtonian flow, the shear rate is proportional to the velocity, and inversely proportional to the distance between the plates. This implies that it has dimensions of velocity divided by distance: in other words, of inverse time: s^{-1}.

Additionally, since the total force acting on the top plate is F, the force acting per unit area of this plate is F/A. This force per unit area is known as the shear stress; and, like other stresses and pressures, has the units of Pascals, Pa. Thus, the dynamic viscosity, η, defined as the ratio of shear stress to shear rate, has the units of Pa s.

5.3 Flow types encountered in paper surface treatment

There are five general flow types encountered in the surface application of chemicals to paper and board.

5.3.1 Newtonian flow

This is the simplest flow regime, in which shear stress, τ, is directly proportional to shear rate, for all levels of shear stress. The viscosity is independent of the shear rate. This is illustrated in Figure 5.2. Flow of this nature governs the behaviour during the surface application of low solids, low viscosity aqueous solutions of water soluble chemicals.

$$\tau = \eta \gamma$$

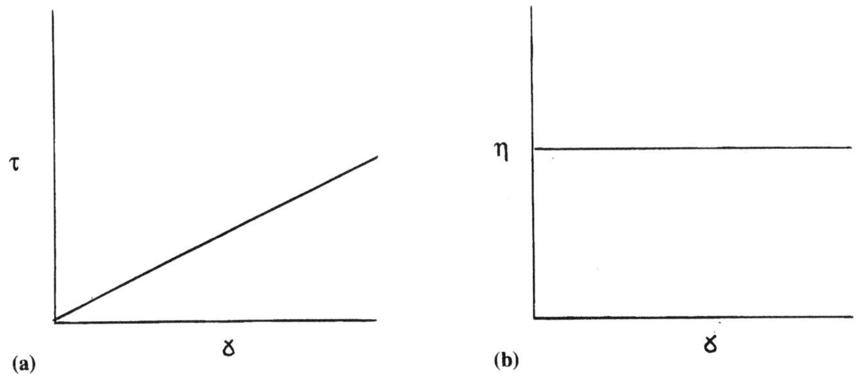

Figure 5.2 Newtonian behaviour. The relationship between (a) shear stress τ and shear rate γ; and (b) shear rate γ and viscosity η.

5.3.2 *Pseudoplastic flow*

Also known as 'shear thinning', this form of flow is characterized by a decrease in viscosity as the shear rate increases. Such behaviour is illustrated in Figure 5.3.

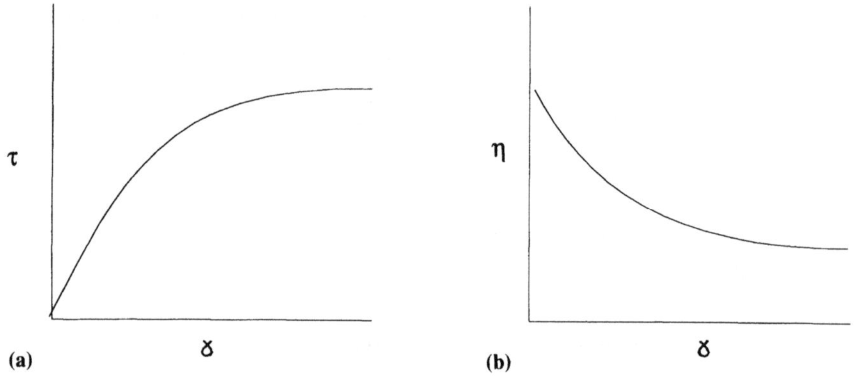

Figure 5.3 Pseudoplastic (shear-thinning) behaviour. The relationship between (a) shear stress τ and shear rate γ; and (b) shear rate γ and viscosity η.

Pseudoplastic flow in surface chemical application is exhibited by high solids suspensions containing high proportions of ultrafine ground calcium carbonate particles.

5.3.3 *Dilatant flow*

This is the reverse of pseudoplastic flow and is thus known also as 'shear thickening'. Its behaviour is illustrated in Figure 5.4.

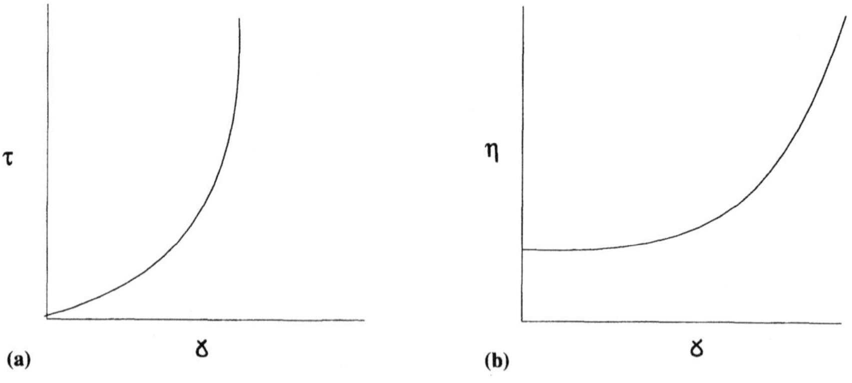

Figure 5.4 Dilatant (shear-thickening) behaviour. The relationship between (a) shear stress τ and shear rate γ; and (b) shear rate γ and viscosity η.

An example of dilatant behaviour is given in Chapter 4, where very high solids starch suspensions are described as exhibiting dilatancy.

5.3.4 *Thixotropic flow*

Strictly speaking, the word 'thixotropic' implies a viscosity which shows time dependence. As such, it can apply to any non-Newtonian flow regime. In practice, it is usually used to indicate time-dependent viscosity loss; fluid viscosity is lost just by exposure to a constant shear rate for an extended time. It can indicate depolymerization of a component in a suspension. It is illustrated by Figure 5.5.

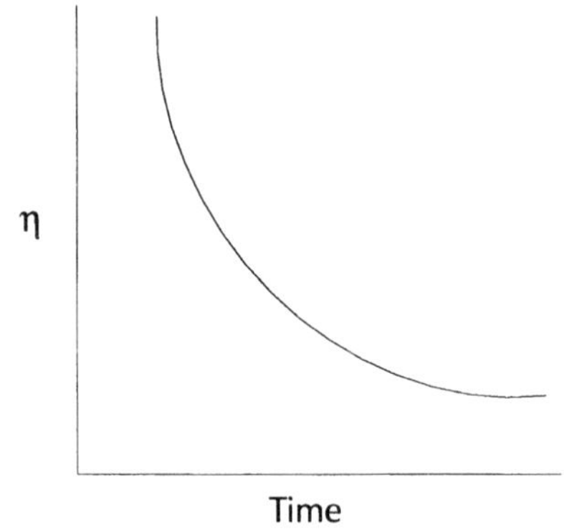

Figure 5.5 Thixotropy: shear-thinning behaviour (decrease in viscosity η) with time, at a constant shear rate.

The best examples of thixotropic flow in the field of surface application are shown by needle-shaped particles, an area which is becoming more important as the use of aragonitic calcium carbonates increases.

5.3.5 *Plastic flow*

The fluid system exhibits a yield value on the shear stress axis which must be exceeded before flow will take place. Examples of this type of flow are rare in surface chemical application because of the obvious practical difficulties that would be encountered. They do, however, occur and some interesting case histories are related.

For example, Newtonian rheology is typically exemplified by low solids coatings (40%–50%) which are applied by air knife and size press. Pseudoplasticity, dilatancy and thixotropy occur in high solids systems

Table 5.2 Rheology requirement for the main coating process

Coating method	Rheological behaviour
Blade, gate roll	Non-Newtonian Pseudoplastic Thixotropic Dilatant
Roll blade	Newtonian/slightly pseudoplastic
Air knife	Newtonian
Metering bar	Newtonian
Size press	Newtonian
Film press	Depends on metering

(58%–70%) where particle–particle interactions strongly influence rheology. Other predominant factors are the different type of pigment, for example, plate-like versus particulate, the ratio of pigments, the particle size of the latex, the viscosity of the starch solution and the type of rheology modifier. High solids systems are usually applied by some form of high shear applicator, for example, blade coater. This generates high shear stresses in the system. Variations in application methods, machine speeds and base-sheet parameters add further influences on the behaviour of the coating colour (Table 5.2).

5.4 Variation of shear stress during surface application

During the overall process of surface application of a chemical from a suspension or solution, the shear stress varies widely. This is illustrated in Figure 5.6. During circulation, from storage to the coating head, the

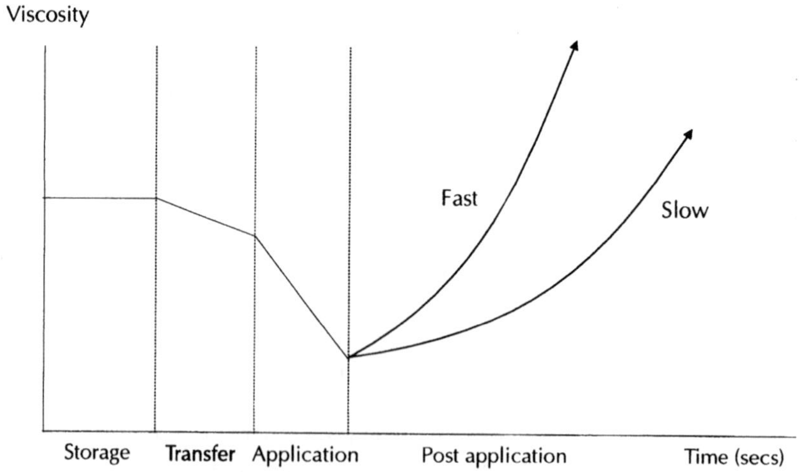

Figure 5.6 Graph of viscosity versus time.

coating colour is subjected to only relatively low shear. This is marked as 'storage' and 'transfer' on Figure 5.6.

At the coating head, particularly where a blade is used for metering, the coating is subjected to very high shear stresses ('application phase' on Figure 5.6).

After the application and metering, all shear forces are removed. This is the levelling phase (post-application on Figure 5.6) when a well-formulated coating will flow sufficiently to form a smooth, level surface.

5.5 Water retention

As their name implies, rheology-modifying chemicals change the flow characteristics of a coating colour. There are many reasons for needing to do this, mainly concerned with the hydrodynamics of the coating process. But there are two very important constraints on available rheology modifiers: water retention and fluorescent whitening agent activation.

Thus, selection of a rheology modifier is a complex issue, entailing studies not only of the coating colour rheology, but also of the interactions with these two constraints.

The water retention characteristic of a coating colour is its ability to maintain an aqueous phase in contact with the pigment and latex particles. This is an essential part of the coating process, as it allows an even build-up of hydrodynamic pressure under the metering blade, leading to even web tensions and smooth coating surfaces. Without a suitable aqueous phase during metering, both coating machine runnability and final product quality can suffer.

These various issues have been widely studied, leading to the development of numerous products more or less effective under different conditions [2, 3].

The following summarizes the main conclusions about water retention.

- Under optimum conditions:
  - the coating colour solids in the application system can be maintained,
  - blade tip pressure runs under a constant load;
  - the fibres do not swell too much;
  - the coating colour on the paper immobilizes quickly enough to minimize binder migration.
- When water retention is below optimum:
  - water is absorbed into the base very quickly, causing rapid setting of the colour on the surface;
  - streaking occurs owing to high solids build-up at the blade tip;
  - there is a high degree of fibre swelling;
  - blade control increases, leading to coat-weight control difficulties and web tension changes.

- When water retention is above optimum:
  - coating penetration is too complete, reducing product opacity;
  - there is more fibre swelling than under optimum conditions;
  - there is a greater chance of binder migration leading to poor printability.

5.6 Types of rheology modifier

There are two major classes of rheology modifier used in the coatings industry: the so-called natural products; and synthetic polymers.

5.6.1 Natural products

The products normally classified as natural products are derived from vegetable materials, and examples are cellulose derivatives, starch, proteins and casein. Carboxymethylcellulose (CMC) derivatives are the most important of this group. Starches are dealt with elsewhere in this publication (Chapter 4).

Carboxymethylcellulose. CMC is derived from cellulose, a natural polysaccharide which is a chain-like molecule consisting of several hundred anhydroglucose units. Each of these units has three reactive hydroxyl groups which can be chemically substituted in a controlled manner such that a wide range of grades is commercially available. The grades differ in their degree of substitution and degree of polymerization. The parameters control the viscosifying property in aqueous media of any CMC grade. Essentially, as the degree of polymerization and carboxyl content increases, the viscosity in water increases and hence the thickening efficiency increases.

When high molecular weight ionic polymers are added to aqueous dispersions containing mineral particles, viscosification is produced by the combination of aqueous phase thickening and particle–particle bridging, which is brought about by the capacity of the polymer chains to adsorb onto the pigment particles by various mechanisms:

- electrostatic – owing to charge differences;
- physical – hydrogen bonding or van der Waals forces.

It is generally accepted that in coating colour formulations the differences in water retention between different grades of CMC are much smaller in carbonate-containing colours than in clay formulations.

This is deemed to be mainly because of the differences in interaction between the CMC and clay particles, which are flat and relatively hydrophobic. This allows a rapid build-up of a network with better water retention properties compared with the reduced interactions which develop with the more crystalline carbonate particles.

Medium viscosity CMC grades are preferred for best water retention for coarse carbonate precoatings whilst low-viscosity grades are preferred for optimum performance in clay-based formulations.

In carbonate formulations, generally higher rates of thickeners are required to achieve optimum viscosity properties.

5.6.2 Synthetic polymers

This group of thickeners is derived from petrochemical sources and most of the available commercial products are manufactured by using the addition polymerization technology of acrylic monomers.

The available chemistry is extensive and the choice of monomers allows for the existence of a wide range of liquid commercial products.

The advantages of liquid products over the solid products derived from vegetable sources encompass:

- ease of handling, transport, storage and dosing;
- non-dusting;
- more resistance to fungal and bacterial attack;
- more flexibility to 'tailor' to a specific end-use requirement.

Although there is only a small number of truly water soluble acrylic monomers available, it is possible to copolymerize in either solution or emulsion form with a number of other monomers such that alkali solubility or swellability can be generated in aqueous media. The lists below indicate some of the possibilities:

- water or alkali soluble monomers:
  - acrylic acid;
  - methacrylic acid;
  - acrylamide;
  - vinyl sulphonic acid;
  - maleic anhydride;
  - vinyl pyrollidone;
- water insoluble co-monomers;
  - acrylic esters;
  - methacrylic esters;
  - styrene.

In addition there is a small number of other functional monomers which are used to produce special properties, for example:

- glycidyl methacrylate;
- di-vinyl benzene;
- *n*-methylol acrylamide.

Specific compositions are proprietary information but extensive patent literature exists.

Polymers may be designed with an essentially linear structure, a branched chain structure or taken to the extreme of developing a three-dimensional network by cross-linking, as illustrated in Figure 5.7. This changes the solubility properties when neutralized more towards swollen spherical particles rather than a truly soluble state.

These polymers are supplied as low-viscosity aqueous emulsions, generally referred to as alkali soluble emulsions (ASEs), in their free acid state (un-ionized). When the pH is increased to between 8–10, the carboxyl groups will ionize causing chain expansion and disentanglement, which results in a significant viscosity increase (Figure 5.8).

Table 5.3 summarizes the effects of molecular weight and cross-linking on thickening efficiency and coating flow properties.

In recent years this basic acrylic chemistry has been further extended by

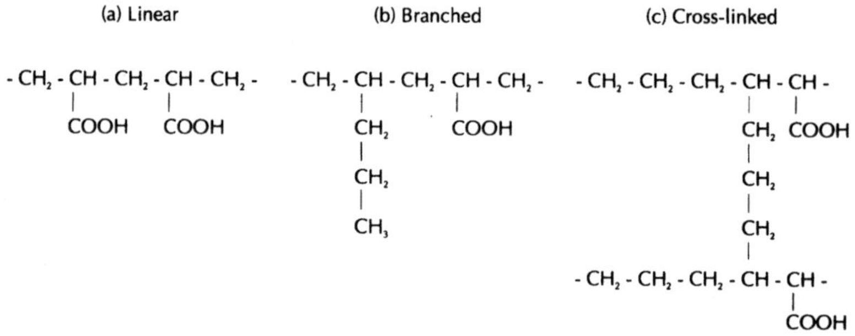

Figure 5.7 Variations in polymer structure: (a) linear; (b) branched; and (c) cross-linked.

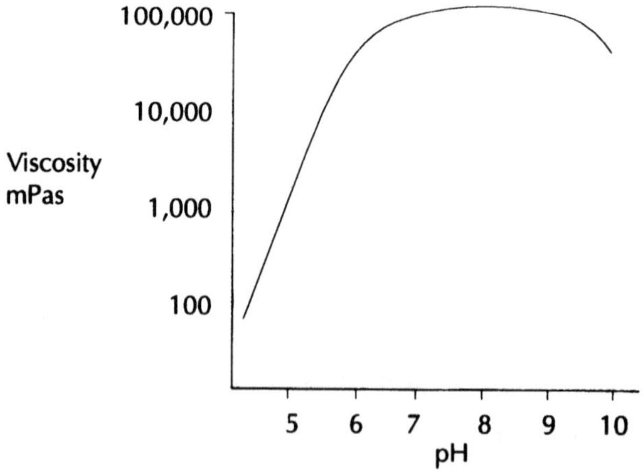

Figure 5.8 Effect on viscosity as pH increases.

Table 5.3 The effect of molecular weight and cross-linking on the performance of carboxylated acrylics

Molecular weight	Cross-linking	Thickening efficiency	Coating flow properties
Low	None	Poor	Good
↓	↓	↓	↑
High	High	Efficient	Poor

the introduction of hydrophobic entities into the polymer backbone, resulting in quite different viscosity and rheology properties.

5.6.3 *Hydrophobic modification of synthetic polymers*

Hydrophobic groups are polymerized into the backbone of the polymer chain such that they are pendent to the backbone along with the carboxyl moiety.

The base variants used to design the backbone polymer are the same in conventional ASEs. This class of thickeners can be described as 'associative' in that the hydrophobic groups associate with emulsion particles present through adsorption onto particles of both emulsion and pigment. In addition, they aggregate together in the water phase in a similar way to that in which surfactants form micelles.

The formation of micelles results in a three-dimensional transient network in which the latex particles serve as the branch points and the thickener molecules act as associative cross-links. These cross-links resist the stretching that must occur when shear is applied to the system. In practice this means improved resistance to shear.

The transient nature of the cross-links, owing to the dynamic equilibrium of the micelle which is continually forming, breaking and re-forming, allows the system to flow under low shear, resulting in excellent levelling.

Because this type of thickener relies on the micelle effect, a critical micelle concentration must be achieved before full thickening is achieved. At normal coating viscosities, this critical micelle concentration will have been passed, allowing the optimum associative action and benefit to be developed.

The entangled polymer chains of conventional acrylic thickeners (Figure 5.9(a)) reorientate themselves along the lines of flow under shear and therefore offer less resistance. With acrylic associative thickeners the entangled polymers are locked together (Figure 5.9(c)) via micelles (Figure 5.9(b)) to give good shear stability and improved resistance to electrolytes.

These associative effects and parameters offer the coating formulator an improved thickening efficiency and modified rheological behaviour. Under high-shear conditions, the micellar structure exhibits a strong tendency to resist disruption.

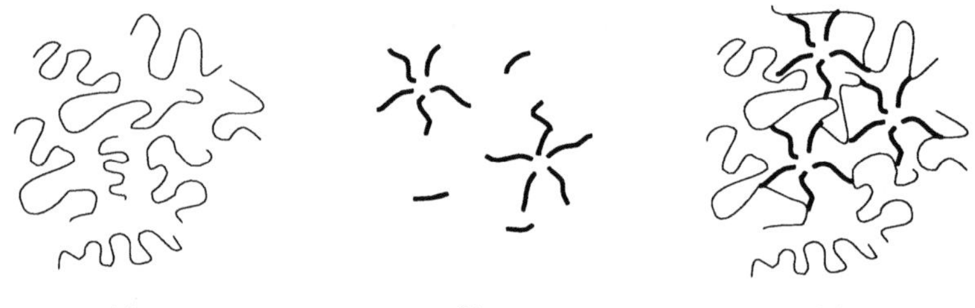

Figure 5.9 Diagrammatic representation of association by hydrophobically modified polymer: (a) solution of water soluble polymer molecules; (b) micelles of non-ionic surfactant; and (c) solution of associated polymer.

Conversely, under low-shear conditions when the coating film is levelling, the state of dynamic equilibrium which exists in the micelles results in a film with good flow-out characteristics. This helps to eliminate streaking and ribbing as the coating begins to dewater and dry.

5.7 Pigment dispersants

Pigments are received at a paper mill either in powder form or as predispersed slurries. The efficiency of the coating process depends upon achieving as high solids content as possible whilst maintaining runnability. Dispersants are used:

- to deflocculate the pigment particles;
- to stabilize the resultant slurry by preventing re-agglomeration (flocculation) of the dispersed particles.

5.7.1 What is dispersion?

The term 'dispersion' is used to describe the process of incorporating very finely divided particles of a solid, typically less than 2 microns particle size, into a fluid (for example, water) at high concentration such that the final product is stable to sedimentation and viscosity variation. The overall process can be summarized as follows:

$$\text{dispersion} = \text{deflocculation} + \text{stabilization}$$

The physicochemical theory of colloidal dispersions is well documented [1–6] and only a cursory treatment will be presented below.

It is considered that there are three stages in the dispersion process;

- wetting of the pigment powder, which also involves the displacement of the air from the internal surfaces between particles in the pigment clusters;

- disruption of the pigment clusters (agglomerates) – usually accomplished mechanically;
- stabilization – the mechanically dispersed particles at high solids content have a natural tendency to re-agglomerate owing to irreversible collision and electrostatic attractive forces, unless a chemical additive, a dispersant, is present.

Pigment particles in water exhibit an isoelectric point which is the pH at which the net charge on the particle surface is zero. Table 5.4 shows the isoelectric points for the common pigments used in paper coating.

It follows then, that as the pH is adjusted away from the isoelectric point, a charge imbalance occurs which leads to particle–particle interaction. These interactions can be minimized by using pigment dispersants.

The classical theory was developed by many workers and culminated in what is known as the DLVO theory – Derjaguin–Landau–Verwey–Overbeek [3, 6] theory.

In practical terms the theory is best explained by reference to kaolin, which has a plate-like structure and has been shown to exhibit a net negative charge on the crystal face and a net positive charge on the edges (Figure 5.10(a)). When dispersed in water at high solids content a 'house-of-cards structure' (Figure 5.10(b)) exists which imbibes the water, thus generating a high viscosity.

Table 5.4 Isoelectric pH values for common pigments

Pigment	pH
TiO_2	4.7
$Al_2Si_2O_5$ (Kaolin)	4.8
ZnO	8.0
$CaCO_3$	9.0

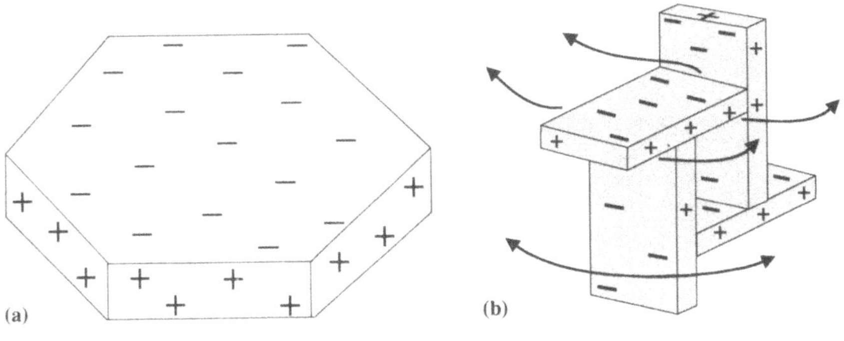

(a) (b)

Figure 5.10 Kaolin: (a) charge distribution on a kaolin platelet; and (b) the 'house-of-cards' structure of kaolin.

The addition of alkali (OH ions) furnishes a preponderance of negative charges on the particle surfaces which generates particle–particle repulsion resulting in the collapse of this structure. The water is freed and the viscosity reduces significantly. This is the deflocculation process.

Stabilization of the deflocculated particles can be achieved by two mechanisms:

- electrostatic repulsion;
- steric or entropic repulsion.

The most common dispersants used in the paper industry utilize the electrostatic repulsion mechanism. Steric stabilization is more commonly used for the organic pigments.

5.7.2 *Electrostatic repulsion theory*

The electrostatic repulsion theory is based on the Helmholtz double-layer approach. Ionic molecules adsorb on the solid particle and impart an ionic charge on the surface (Figure 5.11). As two particles approach, an electrostatic repulsive force is developed, which is a naturally long-range force, resulting in improved stabilization. It has also been suggested that the polyanion (see next section) can act as a protective colloid providing a bumper or steric effect [7].

5.7.3 *Chemical types used as dispersants*

Traditionally, two chemical types have been used:

- inorganic – polyphosphates;
- organic – anionic polyelectrolytes;

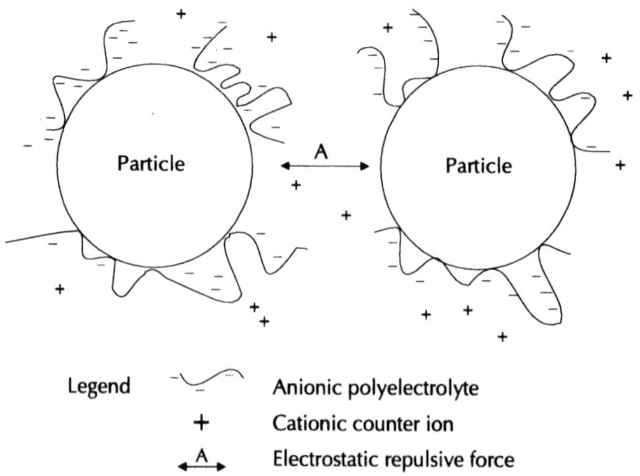

Figure 5.11 Adsorption of polyanions onto pigment particles.

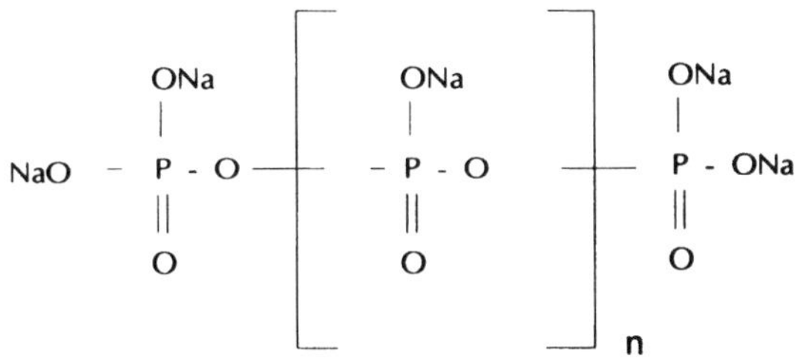

Figure 5.12 Chemical structure of polyphosphate.

but in modern industry polycarboxylates and more specifically polyacrylates predominate.

Inorganic dispersants. Only polyphosphates fit this class (Figure 5.12). In water the molecule ionizes to generate the polyanion. Unfortunately, although polyphosphates are very cost-effective in reducing viscosity and producing the deflocculated state, viscosity instability may occur owing to the poor temperature and hydrolytic stability and to bacterial influences.

Organic dispersants. A wide range of organic chemical compositions is available as pigment dispersants, based on homopolymers and copolymers of:

- maleic anhydride;
- itaconic acid;
- acrylic acid;
- methacrylic acid;

with co-monomers such as:

- styrene;
- isobutylene;
- acrylates;
- methacrylates.

Of the carboxylic acids, only acrylic acid and methacrylic acid are capable of homopolymerization, and, of the two, polymers of acrylic acid (Figure 5.13) are most cost-effective, and therefore it is not surprising that this chemical type predominates.

Dispersant efficiency is assessed by carrying out a dispersant-demand experiment. This is usually undertaken by adding incremental dosages of a dispersant to a high-solids slurry and plotting Brookfield viscosity as the dosage is increased. A typical plot is shown in Figure 5.14.

It should be noted that the collation of comparative data can only be relied upon if a standard method is used. The following factors must be standardized:

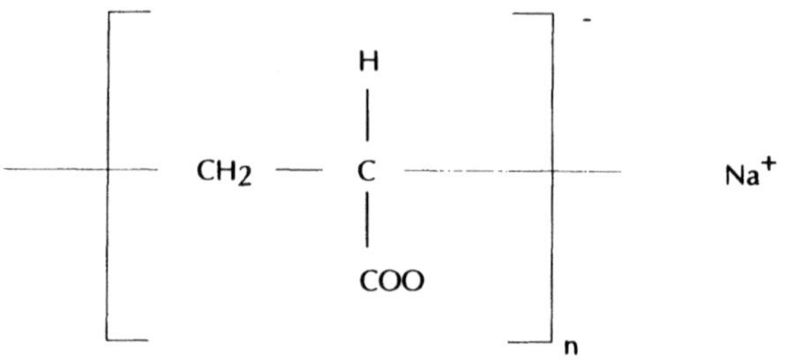

Figure 5.13 Structure of sodium polyphosphate.

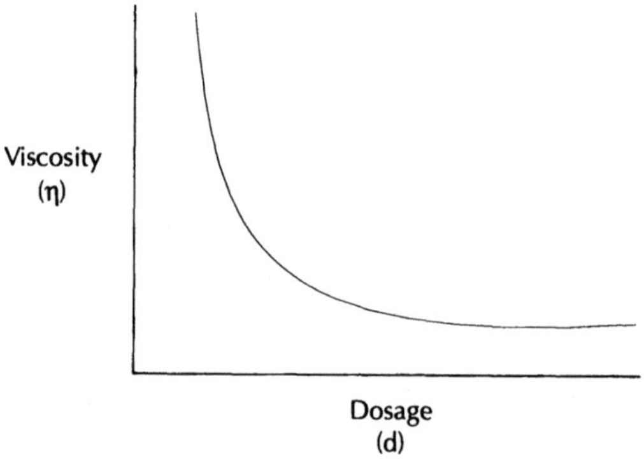

Figure 5.14 Curve of viscosity versus dosage for a typical dispersant; 64% clay slurry.

- particle size and particle size distribution of pigment;
- system temperature and pH;
- water source;
- procedure.

It has been shown by many workers that the effectiveness of pigment dispersants can be influenced by factors such as:

- chemical composition;
- molecular weight and polydispersity.

Figure 5.15 shows the influence of reducing carboxyl content from (a)–(c) on a fixed pigment content slurry, assuming the average molecular weight of the copolymer remains the same.

5.7.4 *Molecular weight effects*

It can be demonstrated that dispersion efficiency is influenced not only by the average molecular weight of a given polymer system but also by its polydispersity.

Figure 5.16 depicts the difference in performance between three polymers of identical chemical composition, which vary in their average molecular weight. As molecular weight increases, so does the dosage d required to produce a minimum viscosity in a slurry of a particular pigment at a given solids content (increase from d_1 to d_3 in Figure 5.16). Furthermore, the minimum viscosity achieved also increases (from η_1 to η_3 in Figure 5.16).

Typical average molecular weights of the commonly used polyacrylates

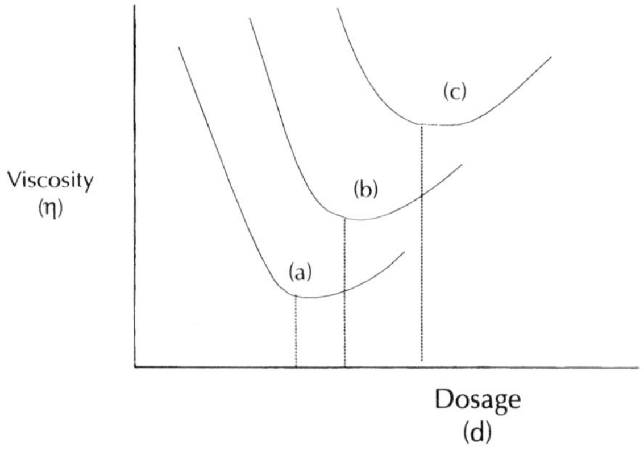

Figure 5.15 The effect of reducing the acid content of the dispersant on the viscosity–dosage curve: (a) highest carboxyl content; (b) 'medium' carboxyl content; and (c) lowest carboxyl content.

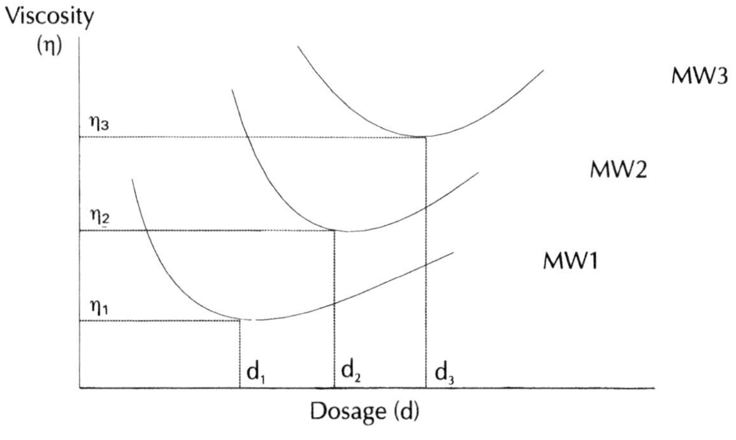

Figure 5.16 The effect of average molecular weight (MW) on the viscosity–dosage curve.

lie in the region 2000–10 000. Thus, given that the molecular weight of sodium polyacrylate is 94, the degree of polymerization is

$$\frac{2000}{94} \simeq 20$$

or

$$\frac{10000}{94} \simeq 100.$$

Most commercial products have a degree of polymerization of around 50.

Polydispersity. However, it is not just the average molecular weight which influences efficiency. The overall polydispersity can be shown to have an influence. Figure 5.17 compares the molecular weight distribution of two polymers of the same average molecular weight. Figure 5.18 shows the relationship between viscosity and dosage with increasing polydispersity, with reference to the polymers illustrated by Figure 5.17.

Dispersants exhibiting low polydispersity are commercially available.

5.8 Interactions

A number of interactions can take place, not only in the pigment dispersion itself but in the finally formulated coating.

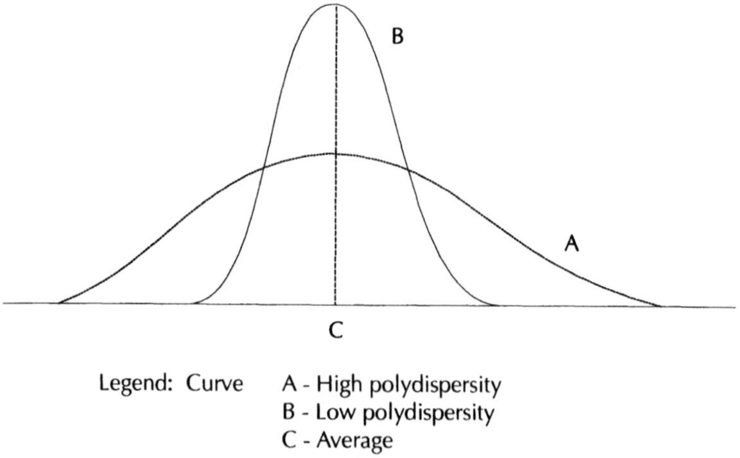

Figure 5.17 Variations in polydispersity of two polymers of the same average molecular weight.

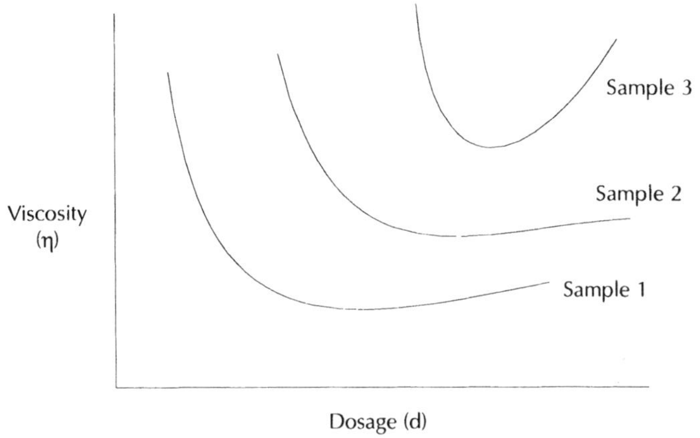

Figure 5.18 The effect of polydispersity on the viscosity–dosage curve: sample 1, low polydispersity (typically, this could be the polymer illustrated by distribution (B) of Figure 5.17); sample 2, 'medium' polydispersity; and sample 3, high polydispersity (typically, this could be the polymer illustrated by distribution (A) of Figure 5.17).

5.8.1 *Effect of pigment particle size distribution*

Differences in particle size distribution occur between different suppliers of the same pigment type. This is because pigment dispersants adsorb on the surface of the particles, and if the particle size distribution contains a higher proportion of fines then the surface area increases substantially, thus increasing the dispersant demand. The converse also applies. This is of particular importance since a higher proportion of fines will be reflected in higher viscosity slurries.

5.8.2 *Effect of multivalent cations*

Polycarboxylates will precipitate in the presence of multivalent cations such as Ca^{2+} and Al^{3+}. The presence of such ions will exert a demand on the pigment dispersant and consequently the dosage required to achieve optimum performance may have to be increased (Schulze–Hardy rule).

Calcium carbonate, of course, has a slight solubility (1500 ppm) but the most common oversight is carrying out laboratory evaluations using either de-ionized water or towns-main and then using recycled or treated bore-hole water in the mill. Laboratory evaluations should include the use of mill water.

5.8.3 *Effect of latex formulation*

In the fully formulated coating there is the potential influence of the latex, it too being a carefully balanced colloidal dispersion. Of course, latex suppliers are aware of the possible interactions and take these factors into acount when finalizing their formulation.

References

1. Parfit, G.D. (1973) *Dispersion of Powders in Liquid*, John Wiley, New York.
2. Patton, T.C. (1974) *Paint Flow and Pigment Dispersion*, John Wiley, New York.
3. Derjaguin, B.V. and Landau, L.D. (1941) *Acta Physicochemica*, **14**.
4. Tadros, T.F. (1987) *Solid/Liquid Dispersions*, Academic Press, London.
5. Goodwin, J.W. (1982) *Colloidal Dispersions*, The Royal Society of Chemistry, London.
6. Verwey, E.J.W. and Overbeek, J.T.G. *Theory of the Stability of Lyophobic Colloids*, Elsevier, Amsterdam.
7. Athey R.D. (1975) *Tappi*, **58** (9), 66–71.

6 Pigments for on-line surface treatment

I. ATTRUP and S. HANSEN

6.1 Introduction

The development in the paper industry has during the past 10–20 years been towards ever higher quality: higher brightness, better opacity, better smoothness and printability.

The traditional way of achieving this has been to coat the paper. High-solids coating colours based on fine clay or calcium carbonate pigments held together by a latex binder are used. The coating colours are usually applied by a blade coater, and coat weights between 8–14 g m^{-2} (sometimes up to 20 g m^{-2}) per side are attained.

Coating is a powerful tool in paper improvement, but it has some drawbacks too. One of the major disadvantages is the relatively high cost.

Surface sizing of the paper with starch is another way to improve the quality. Sizing in itself can improve smoothness and printability, but it has little impact on the optical properties of the paper. It is no wonder, then, that papermakers often have considered adding pigments to the starch solution. It is hoped that adding the pigment, or pigmentizing, as it is called, could prove a quite cheap way of obtaining some improvement in brightness, opacity and smoothness.

The subject of this chapter is pigmentizing. Traditional coating is outside the scope of this book, but today, as we shall see, there is really no firm boundary between coating and pigmentizing.

In the following the terms 'coating colour' and 'coat weight' will be used irrespective of whether we are dealing with size pigmentizing, film coating or traditional coating. The coating colour is the mixture of starch and/or binder and pigments, etc. The coat weight is understood as the weight of pigment plus starch or binder applied per square metre of paper.

For clarification, the following definitions can be made:

- sizing
    - coat weight: 1–2 g m^{-2} per side;
    - medium: starch solution;
    - pigments: none;
    - solids: 8%–12%;
    - application: size press, film press;

- pigmentizing
  - coat weight: 2–4 g m^{-2} per side;
  - main medium: starch;
  - pigments: 20%–40% of total coat weight;
  - solids: 20%–30%;
  - application: size press, film press;
- light coating
  - coat weight: 4–8 g m^{-2} per side;
  - main medium: latex;
  - pigments: 70%–90% of total coat weight;
  - solids: 50%–60%;
  - application: film press;
- traditional coating
  - coat weight: 8–14 g m^{-2} per side;
  - main medium: latex;
  - pigments: 80%–90% of total coat weight;
  - solids: 60%–70%;
  - application: blade coater.

The definitions here have been set somewhat arbitrarily, and examples can be found of applications falling between the categories set up. For pigmentizing, a mixture of starch and latex can be employed, and solids contents or grammages of pigments outside the ranges mentioned here will occur as well. Also, there is nothing to prevent a papermaker from using a blade coater for light coating.

In this chapter, the main subject being pigmentizing, we shall limit ourselves to the lower coat weights, and we have chosen a maximum of 6 g m^{-2} per side, applied either by size press or by film press. We do not, however, differentiate whether the coating colour is based on starch or latex or a mixture of the two.

6.2 The pigments

In the past 10–20 years probably every known type of filler or coating pigment has been tried in connection with pigmentizing.

Most paper pigments today are either based on calcium carbonate or on kaolin clay. The calcium carbonate products can be made from chalk, ground marble or precipitated from milk of lime (PCC).

Chalk is a soft limestone, consisting of fine particles loosely bound together. Paper pigments can be obtained by simple breaking up and separation of the chalk.

Ground calcium carbonate (GCC) is produced by crushing and grinding marble, which is a hard crystalline rock. Some purification, by flotation and magnetic separation is often required.

Precipitated calcium carbonate (PCC) is produced by slaking burnt lime and reacting the hydrated lime with a carbon-dioxide-containing gas, thereby precipitating out the pigment particles directly.

The possible variations in particle shape (morphology) and surface area (measured in m^2 g^{-1}) are not very extensive for chalk or ground marble, whereas there are numerous possibilities with PCC.

There are many different types of clay on the market, from coarse filler grades to fine coating qualities. In many cases the clay needs only little treatment (washing and separation) before pigment fineness is attained. The European natural clay particles are plate-like in shape. In America the clay from nature is less plate-like, but this can be changed by a so-called delamination process. Improved optical properties (scattering) can be obtained by heating the clay to high temperatures (calcining). Calcined clay particles are not plate-like.

Filler clay is normally quite coarse, with a relatively low surface area, whereas the coating grades are finer, with greater surface areas. Calcined clay has an even greater surface area.

Table 6.1 shows typical properties for a number of paper pigments. Brightness is of course an important parameter; PCC will be at the top of the scale, and clay and chalk will have somewhat lower brightness than the ground marble.

More exotic pigments have also been employed, and we shall briefly refer to them later on.

Table 6.1 The pigments calcium carbonate ($CaCO_3$) and clay: their origins and typical properties

Basis, by type	Fineness (percentage < 2 μm)	Specific surface area (m^2 g^{-1})	Brightness (R457)
Filler:			
$CaCO_3$			
chalk	40–60	2–4	80–90
ground	55–75	6–10	85–96
precipitated $CaCO_3$	40–75	4–14	93–98
Clay			
natural	35–55	6–10	78–83
Coating:			
$CaCO_3$			
ground	90–98	11–16	90–95
precipitated $CaCO_3$	85–98	6–12	95–98
Clay			
natural	80–98	12–20	85–90
delaminated	75–95	15–20	85–90
calcined	85–92	19–25	90–92

6.3 Application

Pigmentizing was originally done in the traditional size press. Pigments were added to the size starch, and it proved possible to apply a few grams per side. Control of the process was rather limited, since there was not much to be done by way of adjustments. However, the paper could be made quite uniform in quality and some improvements were obtained. A major problem with pigmentizing is dusting. During many first-time trials with pigmentizing, the reel-up and/or converting equipment would disappear in a cloud of dust. This phenomenon may in some cases be suppressed by adjusting the ratio of starch to pigments, but very likely the problem will then show up at the printers, where milking or deposits during offset printing are observed.

A solution to these problems is to add a proportion of latex to the starch–pigment mixture. The latex will increase the internal bonding in the size and thus tie up the pigment particles better.

With the more modern film press the sizing process is easier to control. Figure 6.1 shows an example of coat-weight control by means of metering rod pressure.

The solids contents of the coating colour can be increased, since the film press allows for higher viscosities. Thus it is possible to apply more pigment per unit area. However, a higher proportion of latex will then be required.

Some machinery manufacturers have further developed the film press in order to apply heavier coat weights (using latex-based high-solids coating colours). The term 'film coater' is sometimes used for these machines.

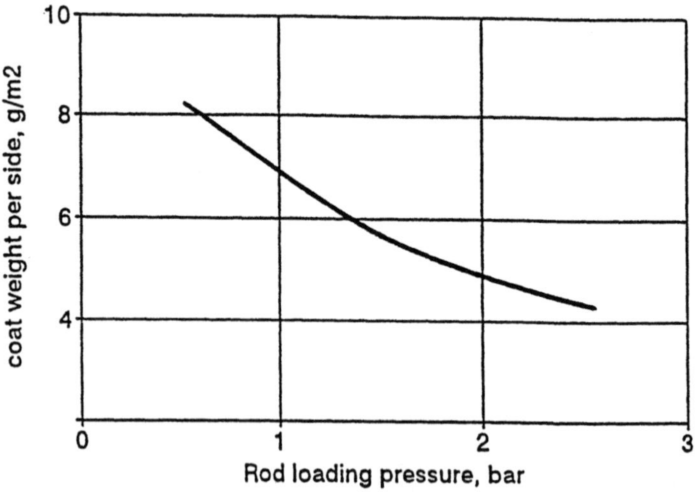

Figure 6.1 Example of coat weight control on a metered size press by adjustment of metering rod pressure.

6.3.1 *Pigment choice*

The choice of pigment from the large selection described above depends to some extent on the purpose of the pigmentizing. This is especially true regarding the pigment fineness. Coarse pigments give a rough paper surface, which is advantageous for later coating. Coarse particles, however, can cause problems with abrasion on the film press rollers.

Abrasion of clay is normally low. Of the calcium carbonate pigments the GCC will give the biggest abrasion problems, while chalk and especially PCC are more favourable.

In most cases the papermaker will prefer the highest possible brightness. Of course, local availability and price of pigments may also have some influence.

Pigmentizing as pre-coat. Pigmentizing is, as mentioned, sometimes employed as a pre-coat for coated paper. It has been seen that various properties of the coated paper can be improved if the base paper has been pigmentized beforehand. For base-paper pigmentizing a rather coarse pigment is usually chosen. Typically, a ground marble product with 60% of particles < 2 µm could be used, but a chalk or a coarse natural clay is also possible.

The main reason for choosing a coarse pigment is that the pigmentized paper should not be too smooth. This is well known from traditional double coating, where coarser pigments are always selected for the pre-coat. The roughness of the precoated paper ensures that the top coat will adhere better, thus giving a better pick strength, more uniformity and better printability.

In one study [1], normal double-coated paper was compared with paper which had been pigmentized prior to the double coating. For pigmentizing, a coating colour was added at 3 g m^{-2} per side. The colour consisted of 100 parts calcium carbonate, 15 parts starch and 7 parts latex, at 42% solids. On the calendered double coated paper, slight improvements in smoothness (PPS) and gloss were observed, while the whiteness and opacity were unchanged.

Another author [2] mentions that by coating on heavily pigmentized paper the following improvements were seen:

- better print quality;
- higher print gloss;
- higher colour intensity;
- better fibre coverage.

6.4 General paper improvement

For uncoated paper, the purpose of pigmentizing is in many cases just to obtain some general improvements in quality. As already mentioned, a higher brightness and opacity is hoped for as well as improved smoothness. Finer pigments are normally employed, e.g., ground marble with 90% of particles < 2 μm, fine PCC, calcined or other coating-grade fine clays.

With 4–6 g m^{-2} at higher coating colour solids, the papermaker is looking for effects similar to those achieved by traditional coating; thus pigmentizing is sometimes looked upon as 'poor man's coating'.

Experience has often shown, however, that 'poor man's coating' is a rather poor coating.

One problem is obviously that it is by no means certain that the pigments actually stay on the paper surface. In a starch-based formulation it is known that both starch and filler tend to migrate into the sheet, to a certain degree. In this case the result of the pigmentizing resembles much more an addition of filler than a coating. A quick calculation shows that this extra amount of pigment is quite limited.

Take, for example, a pigmentizing done on an 80 g m^{-2} paper, at 2 g m^{-2} per side, with 30% pigments and 70% starch. The addition of pigments will then correspond to 2 × 0.6 g m^{-2}, that is, 1.2 g m^{-2}, which is equal to 1.5% extra filler. This extra 'filler' will often be a coating-grade ground marble, which can be expected to be poorly distributed in the sheet and thus not be very efficient regarding brightness and opacity.

In the case where the paper grammage is lowered by the coat weight of the pigment added, in order for the pigmentized paper to maintain the original grammage, the situation is even worse, since some regular filler will be lost in this way. At 20% filler load in the paper, a lowering of the substance by 4 g m^{-2} means that there is 0.8 g m^{-2} less filler in the sheet. Comparing this with the 1.2 g m^{-2} added by the pigmentizing, the net gain in only 0.4 g m^{-2}, and well-distributed filler has been substituted by less efficient pigment. It is no wonder that the result on the opacity may actually be negative.

The situation may be somewhat better with a higher solids latex-based formulation, which more closely resembles a coating colour. Here, it is easier to ensure that the pigments will stay on the surface of the paper. However, even then an application of a small amount, e.g., 2 g m^{-2}, has very little effect on the optical properties. Only when 4–6 g m^{-2} is employed may some benefits be seen.

In one study [3] light coating was attempted on supercalendered paper with a brightness of 70% ISO. The base-paper weight was adjusted so that the coated paper grammage was kept constant at 56 g m^{-2}. Two coating colour formulations were tried, one consisting of 50% coating-grade GCC and 50% high-glossing fine clay; the other formulation was based on 100% delaminated clay. Approximately 12 parts of latex were added, to give a solids content of around 60%. Coat weights of 2, 4 and 6 g m^{-2} per side were tried.

A coat weight of 2 g m^{-2} resulted in a very small gain in paper

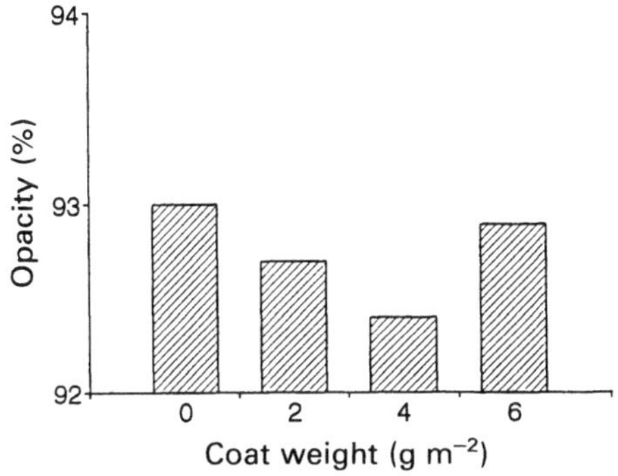

Figure 6.2 Opacity of light-coated super-calendered paper as a function of coat weight and with total paper substance kept constant [3].

brightness, even with the high-brightness mixture of GCC and clay. At 4 g m^{-2} the gain was approximately 1%. At 6 g m^{-2} it was approximately 2%. The coating with 100% delaminated clay fared much worse, giving only 0.8% brightness gain at 6 g m^{-2}.

The results of the coating regarding opacity are shown in Figure 6.1. It is apparent that there is actually a small loss in opacity at 2 and 4 g m^{-2} coat weights, and only at 6 g m^{-2} has the uncoated opacity been attained again. The roughness decreased somewhat at 4 and 6 g m^{-2}.

All told, considering the cost of such a coating compared with the cost of super calendered paper manufacture, it was probably not worth while.

6.4.1 *De-inked-pulp-based paper*

Paper produced from recycled pulp (or de-inked pulp (DIP)) is often rather dark and rather weak. Light coating is more and more frequently employed in order to achieve an acceptable paper quality. Coating by means of the film press is an advantage here.

The mechanical shear force on the paper is far smaller with this method than with blade coating, and a film press is often installed on-line on the paper machine. However, again, a substantial coat weight – at least 5–6 g m^{-2} per side – is required in order to get visible optical results.

6.4.2 *Porosity*

In one particular aspect, pigmentizing has proven to be an extremely powerful tool. By using the right formulations a remarkable reduction in paper porosity can be obtained.

Surface sizing is known to reduce the porosity as measured by, for example, the Gurley method. The Gurley porosity is expressed as the number of seconds it takes for a well-defined volume of air to pass through a standard area of the paper. Thus, what is really measured is the permeability of the paper. A high Gurley figure indicates that air passes through only with difficulty, thus the permeability is low.

As an example let us look at a 110 g m^{-2} paper which unsized has a Gurley value of 7 s. By using starch alone, surface sizing increases the Gurley figure to 12 s. Pigmentizing with an ordinary coating pigment did not significantly increase the value. However, by selecting a high-surface-area product, for example a calcined or a fine delaminated clay (coat weight 2 g m^{-2} per side, 30%–40% pigment) the Gurley figure could be increased to above 20 s. It has been reported that similar results have been achieved

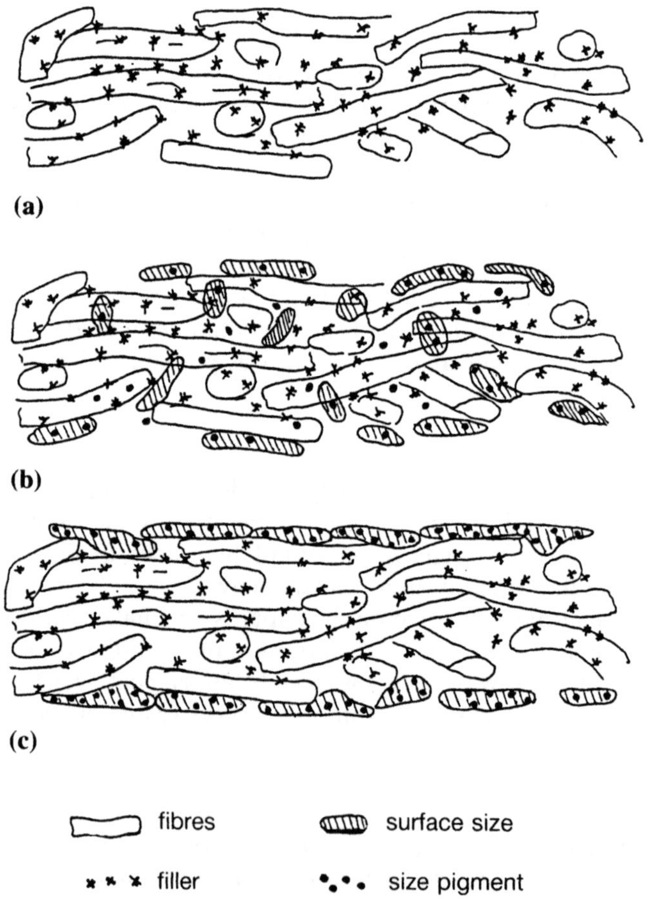

Figure 6.3 The effect of high-surface-area pigments in starch-based pigmentizing: (a) unsized paper; (b) sized paper, with low-surface-area pigment; and (c) sized paper, with high-surface-area pigment.

by using silica pigment with a very high surface area (65 m^2 g^{-1} (private communication)).

In another example, 80 g m^{-2} paper was normally sized at 2.5 g m^{-2} per side to a Gurley porosity of 8 s. By adding a high-surface-area pigment to the size (maintaining 2.5 g m^{-2} with ⅓ pigment and ⅔ cationic starch) the Gurley figure increased to 11 s, corresponding to a drop in porosity of 40%.

A possible mechanism for this startling effect is that the high surface area of the pigment ties up the starch, which is assembled in larger flakes or plates. The plates do not migrate into the paper very easily, but tend to stay on the surface, where they form a more or less unbroken cover or coating. Thus the paper porosity is effectively closed off (Figure 6.3). The low porosity achieved has proven valuable in paper that is to be handled by vacuum, for example, envelopes, cards, etc.

6.5 Conclusions

During the past 10 years there has been considerable interest in pigmentizing and light coating. The emergence of the film press has enabled the papermaker to pigmentize with higher coat weights, and today there is no fixed boundary between proper starch-based low-solids pigmentizing and light coating based on latex and with high solids.

The results gained by pigmentizing and very light coating are so far rather poorly documented in the literature; only a few references can be found. One reason could be that pigmentizing to many papermakers has turned out to be somewhat disappointing. The expected gains in brightness and opacity have not been observed, and severe problems with sheet dusting have been common.

However, in one application pigmentizing has proven its worth. By using high-surface-area pigments a considerable drop in paper porosity can be achieved. Probably this was a discovery by serendipity; the pigments were no doubt employed in the first place because of their light-scattering abilities, stemming from the high surface area.

References

1. Kuchinke, T. (1993) Effect of pigmenting and/or sizing with a pre-metering size press on coated paper quality. Proceedings of the Coating Conference (Minneapolis, MN, 2–6 May, 1993) pp. 151–156 (TAPPI).
2. Turunen, R. (1992) Pigmentizing and soft calendering of printing papers. Proceedings of the Pan-Pacific Pulp Paper Technology Conference, Tokyo, 8–10 September, 1992, part A, pp. 227–33.
3. Burri, P., Drechsel, J. and Balzereit, B. (1995) Pigmenting with metering size press – pilot trial study. Paper presented at the 81st Annual Meeting of CPPA, Montreal, Quebec, January 1995, pp. A19–22.

7 Surface sizing

J. ANDERSON

7.1 Definition and overview

Most definitions of sizing refer to imparting some degree of resistance to the absorption or penetration of liquids, especially water. Applied to surface sizing this implies that the resistance is achieved by applying chemicals to the surface of the paper or board after it has been formed rather than adding chemicals to the wet pulp. In modern technology, surface sizing also involves conferring properties besides hydrophobicity. Parameters loosely referred to as printability, gluability and runnability may also be important in some applications. A better expression than surface size in these circumstances is surface modifier or surface enhancer.

A number of factors have contributed to the resurgence of on-machine surface application of sizing chemicals. There has been a drive towards higher specification paper grades. In some cases the paper properties required would be difficult or impossible to achieve without application of chemicals at the surface. Increased production of coated grades has also encouraged surface sizing which is used to control coating hold-out.

Increased emphasis on environmental issues and the need to clean up the 'wet-end' system have also promoted modern surface sizing. A cleaner wet end allows better control of the papermaking process and makes substantial financial savings possible on large, fast machines. The risk of penalties charged on high levels of chemical oxygen demand (COD) in the mill effluent can provide a strong incentive to increase the amount of chemical treatment at the paper surface where retention of chemicals applied is essentially 100%.

The move towards surface treatment has been made possible, or at least accelerated, by developments in application equipment. These issues have been covered in Chapter 2, but it is important to realize the effect they have had on the use of chemicals at the surface. Introduction of the metered size press has not only allowed surface treatment to be made without reducing the speed of the papermachine, but also changed the way in which the chemicals are applied and where they are located in the finished sheet.

Not surprisingly, with increased application activity at the paper surface there has also been renewed effort by paper chemical suppliers to create surface size materials with properties which confer the appropriate characteristics to the surface and which have good handling properties at the size press.

This chapter traces the development of surface sizing chemicals to their prominent position in modern papermaking technology.

7.2 Chemicals applied at the surface

7.2.1 Types of chemicals

The formulation applied at the size press will usually consist of a mixture of chemicals. These can be classified according to their function:

- transport medium;
- surface strengthening agents;
- speciality chemicals;
- surface sizes or hydrophobing agents.

The surface strengthening agent commonly used is starch. This is the most abundant material present in the formulation, and the starch solution acts as the transport medium or carrier for the surface size. Starches are discussed in Chapter 4 and so no detailed reference is needed here. Nevertheless, it should be stressed that the type, viscosity, temperature and pH of the starch can all affect the amount of surface size applied, where it ends up in the sheet and how it performs. The main sources of starch are potato (farina), maize (corn), wheat and tapioca derived from cassava, which is an important source in hotter regions of the world. The starch is usually a by-product and so its price and purity depend on many factors such as agricultural policy and extraction process. Recently in Europe there has been a reduction in the availability of potato starch and so maize starch is most commonly used. However, wheat starch often offers a cheaper alternative but contains some protein impurities (residual gluten) which can affect foaming properties and viscosity control.

Before it can be used at the size press the native or pearl starch must be converted to modify the viscosity properties of the starch solution. The starch may be preconverted by the starch supplier but most mills prefer to make their own conversion to save costs. The conversion involves breaking bonds in the large starch molecule and can be done by an enzyme, thermomechanical or thermochemical process. The conversion may be a continuous or batch process and selection will depend on cost, viscosity and solids content requirements. Whatever starch and conversion process is chosen it is important to maintain consistent size-press starch solution properties for efficient application of surface size.

The starches may also be modified chemically by reacting other chemical groups onto the converted starch molecule. The most common derivatized starches are ethylated, acetylated, oxidized or have cationic groups added. Acetylated starches give particularly good film-formation properties to the starch. The advantage of cationic starches is that they form ionic bonds with the anionic fibres and fillers and so remain with the fibre during repulping. This has been reported to result in at least a 50% reduction in biological oxygen demand (BOD) [1]. Chemical treatment of the starch increases its cost and so use of chemically-modified starches must usually be justified by giving added value to the paper, improving paper machine runnability or reducing effluent loadings in the backwater.

Although starch is by far the most common additive at the size press, surface size can be applied without starch or with an alternative co-agent, such as polyvinyl alcohol, sodium alginates or carboxymethyl cellulose. These materials are chosen for their good film-forming properties which bond together fibres at the surface and fill holes in the paper structure. Another important property is their viscosity-modifying effect which influences the pick-up of the size-press mixture.

The surface size can also be applied as part of a coating mix which then effectively becomes the carrier. This may be a convenient way to add the size in cases where there is no separate size press available provided the surface size is compatible with the coating mix, but generally the size is less effective when used in this way because it tends to be within the coating layer and less available for influencing the surface properties.

Speciality chemicals can consist of a wide range of chemicals for specific end-use applications. Typically they include materials such as dyes (Chapter 8), fluorescent brightening agents (FBAs) (Chapter 9) and security chemicals.

The surface sizes or hydrophobing agents can be divided conveniently into the following categories:

- rosin-based products;
- alkyl ketene dimer (AKD);
- synthetic polymeric materials;
- others.

Of these the synthetic polymeric materials are the most important in modern surface sizing. They have been developed specifically for application at the surface and are able to produce novel properties in the sheet.

7.2.2 *Wet-end sizes applied at the surface*

Cationic rosin dispersions (CAR) and AKD-based products are wet-end sizes which can also be applied at the surface. The sizing mechanism of CAR at the surface is similar to that in the wet end. Alum or polyaluminium chloride is needed to form flocs of aluminium resinate complex which are deposited on the fibre surface. These flocs melt in the after dryers and react further with the aluminium ions. The cost of CAR emulsions is relatively low compared with synthetic polymeric materials designed specifically for use at the surface, but there are drawbacks. Rosin is an inherently tacky material which gives rise to the risk of deposit formation at the size press or in the after dryers. Cationic rosin is also undesirable if FBAs are included in the size-press formulation. The alum needed for the rosin reaction has a quenching effect on the FBA, making it necessary to increase the addition level. Although CAR can act as a useful surface size it gives surface characteristics limited to those of the rosin molecule and may not be able to meet the performance demands of high-specification paper grades.

AKD can also be applied at the surface as well as at the wet end but the surface addition level is restricted to prevent slip and dusting, as discussed below. Its sizing mechanism has been well described in the literature and involves reaction of the AKD with the cellulose. Unlike the synthetic polymeric materials or CAR, the full sizing effect is not achieved in the after dryers. However, when cured it can confer a high degree of hydrophobicity to the surface. Its use as a surface sizing agent is limited by its waxy nature, its susceptibility to hydrolysis at high temperatures and its delayed curing properties. The same waxy nature that provides hydrophobation also reduces the coefficient of friction of the paper surface and can cause undesirable 'slip' problems during conversion and printing. Hydrolysis of AKD is made more severe in surface applications because it is added to a starch solution which is kept at a high temperature to prevent 'setback'. This affects the starch viscosity and size-press pick-up. Hydrolysed AKD manifests itself as dusting at the paper surface, which has an adverse effect on many printing processes such as offset printing and runnability through a photocopier. Delayed cure can give binder migration when surface sizing is followed by on-line coating.

7.2.3 Synthetic polymeric surface sizes

The synthetic polymeric sizes are of greatest importance to modern surface sizing technology. These materials have been developed specifically for application at the surface. They may be broadly divided into three classes:

- solution polymers made water soluble by forming the anionic salt;
- emulsion polymers consisting of a hydrophobic, water-insoluble polymer surrounded by a protective colloid;
- microdispersions of polyurethane.

The solution polymers are derived from older technology which became popular with the arrival of alkaline papermaking. Traditionally they have been used more widely in North America than in Europe to overcome slip and fugitivity problems associated with misuse of AKD size at the wet end. They consist of sodium, ammonium or mixed salts of styrene copolymers of maleic anhydride, maleic anhydride esters or acrylic acid. The styrene confers the hydrophobic properties to the polymer and the water solubility results from the dissociation of the anionic salt of the other component of the copolymer into solution.

The polymers can be modified by changing the styrene:maleic anhydride ratio, the molecular weight of the polymer, the type of alkali metal salt formed or by the type of alcohol and the chain length of the resulting ester [2]. These parameters can be used to control the solubility of the polymer, its hydrophobicity and its film-forming ability; hence they have an effect on sizing and porosity. Compared with the acrylic emulsions and the polyurethane microdispersions, solution polymers are less hydrophobic. In addition, being essentially 'polymeric soaps' they all share another property which has limited their use at the papermachine: they all generate large amounts of foam which must be checked by addition of defoamer. However, it is important to choose a defoamer type which does not adversely affect the sizing properties of the copolymer. Foam at the size press and in the feed tank is not only an inconvenience but affects pick-up and the stability of the size-press performance.

In common with the anionic salt solution polymers, the emulsion polymers usually consist of a copolymer of styrene to bring hydrophobicity to the sizing system. The hydrophobic core of the emulsion is often, but not always, a styrene–acrylic ester copolymer, and the emulsion polymer surface size can be either anionic or cationic. However, they function differently from the solution polymers and do not have the surfactant properties which cause the generation of high levels of foam. The emulsion polymers have been the most popular group of surface sizes in Europe in recent years, and are becoming increasingly used in North America and other markets producing multipurpose office paper (Chapter 10) [3].

Another class of synthetic polymers used as a surface size is based on polyurethane chemistry. These materials are microdispersions with particle size less than 10 nm. They are low molecular weight polymers to which hydrophobic groups are attached. Many variations can be made in the chemistry of the system including the charge and distribution of hydrophobic groups, but they all create highly hydrophobic surface characteristics. Being microdispersions their behaviour is more similar to solution polymers than to emulsions [4].

7.2.4 Miscellaneous specialities

There are a number of other chemicals which are applied at the surface. These do not fit into any particular class of chemicals but include materials

(a)

(b)

Plate 1 Ink-jet colour bleed: (a) good colour-to-colour edge acuity; and (b) colour-to-colour bleed, showing line growth and high edge roughness.

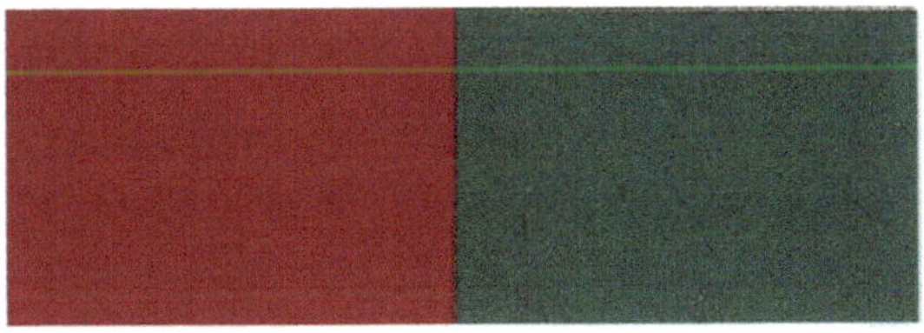

(a)

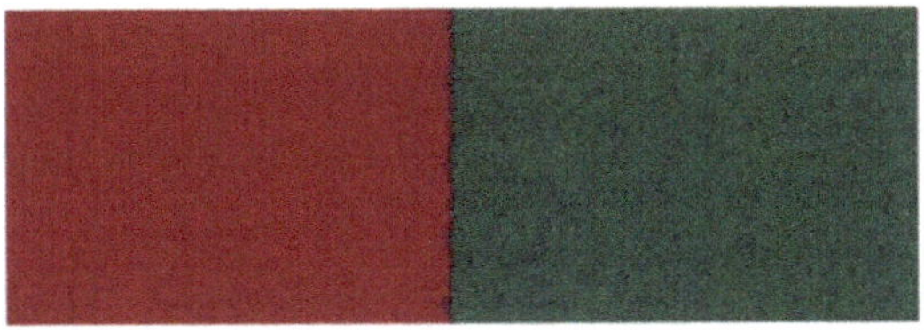

(a)

Plate 2 Ink-jet secondary colour mottle: (a) low mottle; and (b) high mottle.

(a)

(b)

Plate 3 The effect of colour on appearance: (a) blue size-press dyed paper (16 g l^{-1} Cartasol Blue GDF liquid); and (b) yellow size-press dyed paper (16 g l^{-1} Cartasol Yellow 3GSF liquid 150%). The lightness of the yellow dye produces less contrast to the white base paper, therefore appearing to give a less mottled effect

Plate 4 Typical 'orange peel' effect in a size-press dyed paper (20 g l^{-1} Cartasol Brilliant Violet 5BFN liquid).

(a)

(a)

Plate 5 The effect of base-sheet coloration: (a) stock-dyed base sheet (substrate) (0.3% Cartasol Blue K-RL liquid, 0.2% Cartasol Turquoise K-RL liquid); and (b) size-press dying using (a) as substrate (30 g l^{-1} Cartasol Blue K-RL liquid, 30 g l^{-1} Cartasol Turquoise K-RL liquid). By dyeing the stock to a paler colour similar in shade to the final colour required, the contrast of the mottling can be reduced.

such as ammonium zirconium carbonate, stearylated melamine and fluorinated compounds. The use of these materials is restricted to specialist applications.

7.3 Function of the products

There are important differences in the way the synthetic polymeric surface sizes function, the properties they confer to the paper or board, and in their compatibility and handling characteristics.

7.3.1 Solution polymers

These are all anionically charged polymers and so cannot be used at low pH conditions because the dissociated carboxyl groups which provide the water solubility of the polymer are protonated and the solubility is lost. They function by using a cationic species to 'bridge' the anionically charged polymer to the anionic fibre surface. As with cationic rosin emulsions alum is commonly used to provide a multivalent cation to which the carboxylate ions of the polymer salt can coordinate and thereby attach to the paper sheet. The hydrophobic styrene groups are directed away from the surface, and sizing is developed as seen by a decrease in Cobb value or an increase in Hercules Size Tester (HST) value. A higher proportion of styrene in the copolymer gives greater sizing response.

In addition to introducing hydrophobic groups the solution polymers also have film-forming ability which affects the porosity of the sheet and can give an increase in coefficient of friction compared with the base sheet. The molecular weight of the polymer influences both its sizing ability and its film-forming properties. An increase in the degree of polymerization increases the sizing response and improves the film-forming ability. In styrene–maleic anhydride copolymers the molecular weight can also be changed by forming an ester with the carbonyl group from the maleic anhydride. The type of cation used to form the salt is also claimed to influence the properties of the polymer. The ammonium salt gives better sizing response because it is easier to dissociate, but the sodium salt has better film-forming ability.

There is a wide range of styrene–maleic anhydride (SMA), SMA ester and styrene–acrylic acid copolymer (SAA or SAR) solution polymers available. The choice of product used depends on many factors, including the degree of hydrophobicity required, film-forming properties, solubility and active solids content of the polymer solution, foaming tendency and tolerance to salt added at the size press to improve surface conductivity and reduce the build-up of static charge.

7.3.2 Acrylic emulsion polymers

The acrylic emulsion polymers may be either anionically or cationically charged. The cationic emulsion products usually contain a cationic polymer which has greater ability to attach to the fibre and so can be more efficient in systems with a high concentration of anionic impurities.

In a similar way to the solution polymers, the anionic emulsions prefer some cationic species to be present to bridge to the fibre. The requirement is much less for the emulsion products which are able to operate over a much wider pH range because of their different sizing mechanism. The anionic products are, in fact, only mildly anionic or even amphoteric in nature. The polymer particles in the emulsion are typically 50–100 nm in diameter and attach to the fibre by hydrogen bonds. The main difference from the solution polymers is that the emulsion polymer particles flow across the surface of the sheet when they are heated in the after dryers. This affects their distribution across the surface and penetration into the sheet. The chemical composition of the acrylic copolymer can be altered to change its melting or glass transition temperature, thereby influencing the way it flows and spreads in the after dryers.

The emulsion polymers usually have a better compatibility than the solution polymers with other materials added in the size press. They have some film-forming ability but less so than the SMA and SAA products. Because the anionic emulsion polymers are not surface active they are low-foaming.

7.3.3 Polyurethane microdispersions

The polyurethane polymers consist of a low molecular weight polymeric backbone along which are positioned hydrophobic tails. The cationic polymer can attach directly to the anionic groups on the fibre, but the anionic polymer requires a cationic 'bridge'. The bridge is formed typically by alum or cationic starch applied at the wet end. An important feature of the polyurethane structure is that the hydrophobic groups can rotate on their polyurethane axis when heat is applied in the after dryers. The hydrophilic groups on the polymer then align with the fibre, and the hydrophobic groups orientate away from the surface, thereby giving a much higher degree of sizing than the syrene–copolymer solution. However, unlike AKD, sizing is developed in the dryers, the coefficient of friction is not reduced and the polymer is not hydrolysed to create dusting. The polyurethane microdispersions are not film-forming but provide a flexible, hydrophobic barrier which has good resistance to extremes of pH. This makes them particularly suitable for use in one-side coated applications as discussed in the next section.

7.4 End-use properties

7.4.1 Typical end-use properties

The size press provides the last opportunity to modify the chemical properties of paper, or prepare it for coating. The properties to be delivered by the surface size vary greatly and depend on the end use for which the paper and board is intended. In some cases the precise physical and chemical properties required are not well defined, and the end-use property is specified only by empirical performance tests such as ink flotation or dust generation after a prolonged period.

Most of the end-use properties that the surface size should deliver will include one or several of the following:

- increased hydrophobicity;
- film-formation or barrier properties;
- increased surface strength;
- modified frictional properties;
- modified optical properties;
- reduced sheet porosity.

As the surface size is normally used with a carrier, the properties of the surface size are not discrete and interactions occur. Two of the more important properties for which the surface size is responsible are hydrophobation and barrier formation.

As discussed in the previous section, the hydrophobic groups most commonly used in surface sizes are long-chain fatty acids and phenyl groups. Long-chain fatty acids are the familiar hydrophobe in AKD wet-end sizes. They are also present in the polyurethane microdispersions where they can rotate on the polymer backbone and orientate towards the surface as explained in Section 7.3.3. With the solution and emulsion polymers phenyl groups derived from the styrene copolymer introduce the hydrophobicity. In the rosin-based surface sizes abietic acid is used in a similar way to wet-end sizing. The important feature about surface sizing with synthetic polymers is that they provide 'instant' hydrophobation with essentially 100% retention of size.

True film-formation is developed by the copolymers containing carboxyl groups capable of hydrogen bonding to form films. However, a type of 'film' can also be introduced by a polymer distributed across the surface which flows in the after dryers to form a barrier to the penetration of oils and grease and which also limits the migration to the surface of materials internally in the sheet. In the case of coating, the objective is often to limit the migration of binders or microcapsules into the sheet.

Because surface sizes are used over a wide range of end-use applications it is necessary to translate physical chemical properties such as hydrophobation and film-formation into end-use performance-related properties such as printability in fine paper, coating hold-out and gluability in recycled grades.

7.4.2 *Plain fine paper*

Plain fine paper consists of uncoated grades usually made from virgin fibre. A large part of this paper segment is multipurpose office paper or copier paper which is produced in a range of qualities. Apart from some hydrophobicity the main requirements of the sheet are appearance, especially brightness, smoothness and stiffness, printability, usually on a range of printing systems, and runnability through printing machines and photocopiers. These requirements have become more demanding in recent years with the trend towards brighter and smoother paper, the increased use of ink-jet printing, the arrival of faster printing and copier machines, and in some cases the introduction of recycled fibre into the sheet. As in most papermaking activities the end result is usually a compromise or trade off of properties and treatments to produce an acceptable end result.

Surface sizing is increasingly used in combination with some internal sizing to achieve the exacting end-use properties required. These include:

- black text wicking and feathering in ink-jet print (IJP);
- IJP colour-to-colour bleed;
- IJP print density;
- toner adhesion for copiers and laser printers;
- sheet feed through copiers;
- linting and dusting in offset printing.

The composition of fine paper varies widely in terms of filler and fibre type and content, internal size, retention system and machine type and so it is difficult to generalize about optimum conditions. A typical European fine paper might consist of:

- a mixture of long and short fibre, sometimes including eucalyptus;
- neutral or alkaline sizing with AKD or alkenyl succinic anhydride (ASA);
- 15%–20% ground calcium carbonate or precipitated calcium carbonate filler;
- fluorescent brightening agent (FBA) added internally and at the surface;
- starch, surface size and probably some salt added at the surface.

The choice of starch used is often decided by cost and availability rather than by performance. Enzyme-converted maize starch is commonly used, although there are some examples of low cationic starch and ethoxylated starch in use. Because a fluorescent brightening agent (FBA) is usually included in the size-press mixture an anionic surface size is preferred. In Europe the styrene–acrylate emulsion polymers are most common but in North America styrene–maleic anhydride or styrene–acrylic solution polymers are in use.

A typical sizing specification for a plain fine paper is a Cobb(60 s) value of 20–25 g m^{-2} or a HST value (80% reflectance, 1% formic acid, naphthol green ink) of 200–300 s, although examples can be found of harder and softer sized sheets. The HST value is more sensitive to the effects of the surface treatment. The surface size usually contributes a relatively small amount to the sizing property but can have a large effect on the printability and runnability properties. This makes is difficult to discuss surface sizes in conventional papermaking terms.

The requirements and behaviour of the ink-jet systems are particularly difficult to explain in these terms. These systems involve a large number of very small alkaline ink droplets being fired very rapidly (1 droplet every 20–30 μs) at the paper sheet or 'substrate' at which point the definition and quality of the image is determined. The inks are water-based but contain a range of surfactants and cosolvents to control the wetting, spreading and penetration into the sheet. The inks are commonly dye-based but black pigment inks are now used in some of the current printer systems. The aim is to produce an image with strong, bright colours (high print density), good edge acuity (low feathering, wicking and colour bleed), even print distribution over heavily printed colour areas (low mottle), an ink that dries quickly preventing set-off to the next sheet (short dry time) and which does not pass through to the reverse side of the sheet (low 'show-through' or 'strike-through'). Examples of these effects are shown in Figures 7.1–7.3 and Plates 1 and 2. Added to this, most printer systems have different ink formulations, printer head architecture and colour management software systems. It is not surprising then that measurements of Cobb, HST and air porosity of the sheet have failed to predict IJP performance.

The print–paper interactions are obviously complex and not fully understood, which is why a large amount of research and development is in progress to understand them better and improve print quality in the existing and new printer systems which appear on the market each year. Application of starch at the surface usually contributes towards increasing the print density but is not sufficient to give optimum print definition. A surface size such as a styrene acrylate emulsion added typically at 1–5 kg t^{-1} of 25%–35% solids emulsion on paper improves the print quality by modifying the paper-wetting, spreading and absorption properties. These are related to the surface free energy and contact angle of the sheet. Some of the equipment recently introduced to measure dynamic changes in penetration, absorption and contact angle offer better information about the ink-jet behaviour since they can be used with ink-jet inks and are able to measure changes in properties over very short time periods.

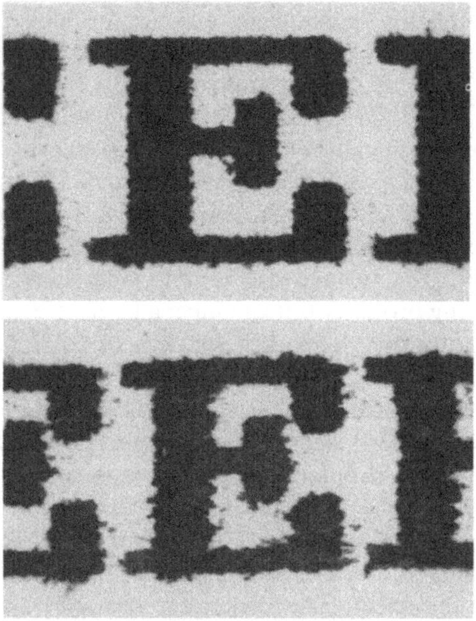

(a)

(b)

Figure 7.1 Ink-jet black text feathering: (a) good black text edge acuity; and (b) black text feathering.

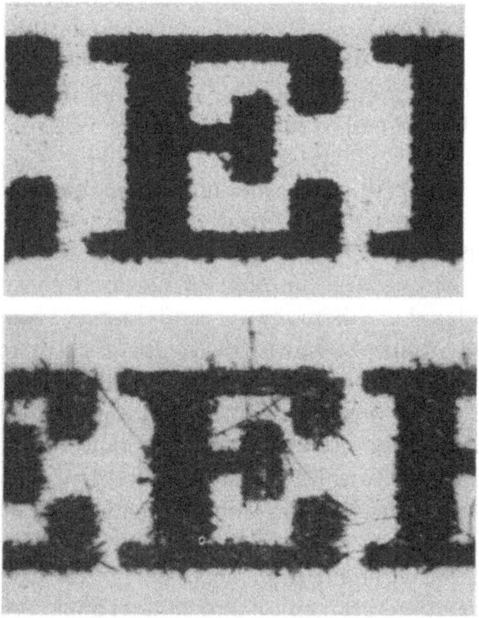

(a)

(b)

Figure 7.2 Ink-jet black text wicking: (a) good black text edge acuity; and (b) black text wicking along the fibres.

Figure 7.3 Ink-jet print show-through: (a) no print show-through; and (b) bad print show-through.

Future trends are expected to include faster ink-jet printing systems for the small and home office user. This will mean that a short drying time will become more critical. Many attempts have been made to include a small amount (typically 3–5 g m^{-2}) of pigment in the size-press mixture to increase the absorption capacity at the surface and increase the drying rate. However, care must be taken not to reduce print density by this treatment.

As noted at the beginning of this section, multipurpose office paper must be able to perform well on a range of impact and non-impact printing systems. It is not sufficient for the paper to give good IJP quality if it is unable to pass well through a photocopier or has poor toner adhesion properties. Toners are dry powders consisting of mainly resin and pigment (carbon black), although there are small amounts of speciality materials which are unique to the particular printer system. Black toners usually contain a styrene–acrylic polymer, and so a styrene acrylate surface size, because of its similar chemical properties, usually gives improved toner adhesion properties. Three types of test are commonly used to assess toner adhesion. One test is referred to as a 'tape-peel' test in which the paper sample is printed with a toner ink (either copier or laser printer), an adhesive tape is applied to the printed area and then peeled off under controlled conditions. The ratio of the print density after tape peel to the original print density gives the retained print density (RPD); the higher the RPD the better the toner adhesion. In another test the paper sample is printed and then folded to simulate a letter being inserted into an envelope. The width of the crack along the print gives an indication of the toner adhesion: the greater the width the lower the toner adhesion. The third test involves rubbing a paper surface a number of times across a toner printed area and either measuring the mass of toner removed or the decrease in print density of the printed area.

The requirements of offset printing must also be satisfied. A concern of offset printing is the build-up of dust from the paper sheet on the printing blanket. There seems to be no satisfactory test method available to measure and predict this behaviour other than to print a large number of sheets until the build up of dust makes it necessary to stop the printer and clean the blanket. Styrene acrylate emulsion polymers and polyurethane microdispersion surface sizes have both been reported to reduce dusting and linting.

7.4.3 Coated paper and specialities

As the name suggests, specialities include a wide range of diverse paper types which often form niche market segments. One common feature of these grades is the use of a surface size to form a barrier on one side of the sheet. Its function may be to stop the penetration of a compound of the coating mix from passing through the sheet, as in the case of the microcapsules applied in the coating mix in the production of carbonless copy paper; or to stop glue passing to the reverse side as in the the case of one-side coated labels. Surface sizing is the preferred treatment because the size is placed at the surface where it is required and the sizing effect is developed 'instantly'; unlike AKD size which requires a time to cure. This is important for on-line coating since the coating treatment then follows directly after the surface size. In the example of high solids coating mixes the sizing is required prior to coating to prevent the water from the coating mix penetrating into the sheet and preventing an even application of coating on the surface.

The choice of surface size depends very much on the coating treatment and the end-use application. If very hard sizing (Cobb(60 s) < 10 g m^{-2}) is required, a polyurethane surface size can be used without risk of the dusting problems associated with AKD applied at the surface; this can also present a high concentration of hydrophobic groups at the surface. The polyurethanes are also flexible polymers, which allows them to resist 'cracking' or 'breaking' during hard calendering. They are also resistant to extremes of pH once applied on the sheet, which is important in some post-treatments such as development of diazo papers under highly alkaline conditions.

One drawback of the highly hydrophobic surface of the polyurethanes may be that it reduces the adhesion of glues and adhesives to the sheet. The styrene acrylates are less hydrophobic and have good gluing properties. Addition levels of surface size vary greatly in this market segment. If a particular specification has to be met then a higher surface size addition level may be justified in a speciality market segment.

7.4.4 Recycled grades

These include paper and board made from recycled fibre. By far the largest market segment is the test-liner and fluting segment where mechanical fibre, typically old corrugated containers, are recycled without de-inking or an extensive cleaning procedure. This means the board is often produced under conditions contaminated by a high concentration of anionic trash. In extreme cases internal sizing may be completely ineffective.

The properties needed in this segment usually include:

- long-term hydrophobicity (low Cobb(30 min));
- good gluability of hot melts, cold adhesives or tapes;
- satisfactory printability on some surfaces.

Because of the prevailing anionic conditions cationic surface sizes are usually used because of their excellent sizing efficiency. Cationic styrene-acrylate emulsion products are popular, often used with alum or polyaluminium chloride.

Recycled grades may also include recycled writing and printing papers in which the fibres are a mixture of chemical and mechanical types and some de-inking may have been performed. A particular characteristic of this grade is the high level of fines in the stock and the variable amount and type of filler present. Again, internal sizing may be difficult to achieve and the variable composition of the stock makes it difficult to control. However, the main concern is often dusting at the surface and wicking of print into the fibres because of the fines and broken fibres present. In this case a surface size is essential to improve the surface properties and a styrene acrylate emulsion or styrene–maleic anhydride solution polymer may be used.

7.5 Interactions and possible problems

7.5.1 Chemical compatibility at the size press

A prerequisite for application of a surface size at the size press is that the size should be in a satisfactory condition for optimum sizing efficiency when it is applied at the surface. When first added to the size-press solution it should combine with the other chemicals present to form a free-flowing homogeneous mixture without any separation or precipitation of any of the components. Once present in the mixture it will make a number of passes around the size-press feed loop because a proportion of the solution entering the size press will be recirculated back to the feed tank. The proportion that is returned will depend on many factors related to the size press pick-up, but especially to the type of size press used. There will be much higher recirculation in a film-transfer size press than in a conventional puddle size press. The capacity of the feed loop to the size press, the volume of the feed tank and, indeed, the addition point of the surface size will all affect the life experience of the surface size before it finally comes to rest at the paper surface. It is recommended that the surface size is added as close to the size press as possible and that the final feed loop is as small as possible to minimize detrimental effects on the size-press solution. Nevertheless, it is still wise to check the initial compatibility of the surface size in the size-press solution by adding the size to a small sample of the solution at an addition level similar or slightly higher than expected in mill use and to observe any signs of separation or inhomogeneity in the mixture after several minutes standing. If there is evidence of incompatibility it may be necessary to change the solution conditions or to select an alternative surface size, or both.

Properties of the size-press solution which have most effect on surface size compatibility are pH and the presence of bivalent metal ions. Starch, salt and fluorescent brightening agent are the other chemicals normally present in the mixture and can affect compatibility and performance. Naturally, the cationic surface sizes cannot be used at alkaline pH or with FBAs. Similarly the anionic products have a minimum pH at which they can be applied. The anionic solution polymers are particularly sensitive to pH because they must remain as the anionic salt if they are to function as a surface size. They are also sensitive to the presence of salt.

7.5.2 Foam

Foam can affect many aspects of the papermaking process and surface sizing is no exception. Some of the surface sizing products inherently have foaming properties. The solution polymers may be regarded as polymeric soaps and so they are expecially prone to the generation of foam. With these surface sizes addition of a defoamer to the starch tank or closer to the size press is usually essential. Foam can also build up during the recirculation process, especially with some types of pump; or if the recirculated solution is allowed to drop back into the holding tank from a level above the surface. Naturally, it is prudent to engineer out as many potential points of foam generation as possible.

7.5.3 Contamination in closed systems

In paper mills with very closed water loops, contaminated water from the wet end can pass through to the size-press solution via recirculation of the starch solution at the size press. This is most common in systems using recycled fibre such as mills making liner board. In these cases the conductivity in the backwater can increase to high levels from the build up of anionic trash extracted from the recycled fibre. Internal sizing may then be ineffective and a surface size is essential. Often a cationic surface size is preferred because of its higher sizing efficiency, but with recirculation of the size-press solution and contamination from the wet end the efficiency can be decreased. An anionic trash catcher such as polyaluminium chloride can be helpful to reduce the build up of the anionic impurities and maintain good surface sizing.

7.5.4 *Size-press pick-up*

Apart from the chemical structure of the surface size molecule, the function of the size is affected by parameters at the wet end, in the size press and in the after dryers and post-treatment. The following interactions affect the pick-up of the size-press formulation and the final location of the surface size in the finished sheet:

- type of surface applicator (film-transfer or puddle size press);
- moisture content and degree of internal size of the sheet entering the size press;
- temperature, viscosity and concentration of the size-press formulation, extent of recirculation and foam formation;
- mechanical conditions at the size press (for example, nip pressure, type of roll) and papermachine speed;
- conditions in the size-press after dryers (for example, drying time and temperature profile);
- post-treatments such as calendering and off-machine coating.

The amount of surface size applied to the sheet depends on a combination of the concentration of surface size in the size-press solution and on the size-press pick-up. The moisture content of the sheet entering the size press and the level of presizing before the press have the greatest effect on the pick-up and the level of sizing achieved.

7.5.5 *Fixation and distribution of the size*

Not only is the amount of surface size applied important but also where it ends up on the surface, how it is distributed within the starch and whether there is penetration of the surface size into the sheet. The anionic surface sizes are fixed to the fibre by hydrogen bonds or by cationic bridges formed by alum or polymers and so this mechanism will also affect the distribution. Some surface sizes such as the solution polymers have surface active properties, whereas the emulsion polymers do not. The surface active properties are responsible for the tendency to foam but also affect the distribution of the surface size within the starch. The distribution of the surface size on and in the paper sheet is the subject of current research.

7.5.6 *Influence of post-treatment*

After the surface size has been applied at the size press it normally passes through after dryers; this may be followed by calendering or coating treatments. The drying regime affects the final form of the surface size, and if the profile or type of heating is inappropriate for the surface size applied there can be problems from deposit formation at the dryers. A higher concentration of size-press chemicals will be applied using a film-transfer size press and these will penetrate less into the sheet. Often infra-red dryers are used in this case since there is less water to be removed and there is less risk of deposit formation. The softening point or the glass transition temperature of the emulsion polymers can be modified according to drying conditions. For instance, with a film-transfer size press followed by infra-red post-drying, a low softening point polymer may be preferred to increase the spreading of the polymer across the surface of the sheet and give a more even coverage of the polymer. Conversely, with a conventional size press followed by contact drying to remove the larger amount of water present it may be better to use a higher softening point polymer to avoid the risk of deposit formation at the drying cylinders.

For polyurethanes, drying conditions are important for other reasons. Sufficient heat is required to allow the hydrophobic groups to reorientate and develop a high level of hydrophobicity. However, lower temperatures than for the emulsion polymers are usually sufficient because no melting and flowing mechanism is involved.

In the calendering process a lot of energy and high pressures are applied to the sheet. The flexibility of the polymer determines how the polymer responds to these conditions. Polyurethanes are considered to be a more flexible polymer and less prone to changes from the pressure and energy applied.

7.5.7 Balanced sizing

In this chapter surface sizing has been put under the spotlight but we should remember that it is only one of the steps, albeit an important one, in the overall papermaking process. The final properties of the paper and board are influenced by the treatment given at the wet end, the surface and in post-treatments. Although surface sizing gives essentially 100% retention the effect of the surface treatment still depends on the properties of the base paper entering the size press. Where the end-use application of the paper requires good printability, good base-paper properties are essential to give a good finished product. Surface sizing should not be regarded as a miracle cure or a way of covering up deficiencies at the wet end. The most effective treatment usually involves a good balance between the wet end and the surface treatment. Combined internal and surface sizing is recommended to produce optimum paper properties in a clean and cost-effective system [5].

References

1. McQueary, R.T. and Thomas, T.W. (1991) *TAPPI Proceedings of the Papermakers Conference*, p. 185.
2. Batten, G.L. Jr, (1992) *TAPPI Proceedings of the Papermakers Conference*, p. 159.
3. Tsai, Y.-G., Colasurdo, A.R. and Cordoba, C. (1995) *TAPPI Proceedings of the Papermakers Conference*, p. 427.
4. Moore, G. (1985) *Paper*, 9 September, p. 22.
5. Tompkins, T.W. and Shepler, J.A. (1991) *TAPPI Proceedings of the Papermakers Conference*, p. 191.

8 Surface brightening of paper

H. WEAVER

8.1 Introduction

White pulp has an intrinsic yellow hue caused by increased light absorption in the blue part of the visible spectrum compared with the yellow/red part of the spectrum. (The visible spectrum spans the range 400–700 nm, from violet at 400 nm through blue, green, yellow, orange, to red at 700 nm.)

It can be seen in Figure 8.1(a) that white pulp reflects to a significantly greater extent in the yellow/red part of the spectrum. This 'blue deficit' in the reflectance curve has been traditionally compensated for by the addition of blue shading dye or pigment, e.g., ultramarine.

Ultramarine actually absorbs in the yellow/red region and reduces the yellow reflectance of the pulp (Figure 8.1(b)). The use of dyes such as this also reduces the total light reflected by the pulp, making it seem very slightly dull and grey. A more effective way of reducing the yellowness of pulp would be to increase reflectance in the blue visible part of the spectrum.

A new group of compounds was introduced in the 1950s which improved the brilliance of pulp by increasing blue reflectance. The group became known colloquially as 'optical brightening agents' (OBAs) or 'fluorescent whitening agents' (FWAs) and quickly became commercially important. They are, however, better described as 'fluorescent brightening agents' (FBAs).

FBAs absorb ultraviolet light in the range of 350–400 nm (near ultraviolet) and re-emit it (fluoresce) in the range 400–500 nm, with peak reflectance in the range of 435–445 nm (Figure 8.2).

FBAs are now important constituents of most printing and writing papers, both coated and uncoated, and of many other grades as well. They are generally easy to use, provided certain precautions are taken. They are relatively inexpensive and may be added either to the body of the paper (stock addition) or to its surface (surface addition).

8.2 Spectra of different illuminants

Fortunately, sunlight has a proportion of its light in the range 350–400 nm, though very little below 350 nm (Figure 8.3). Artificial lights vary in

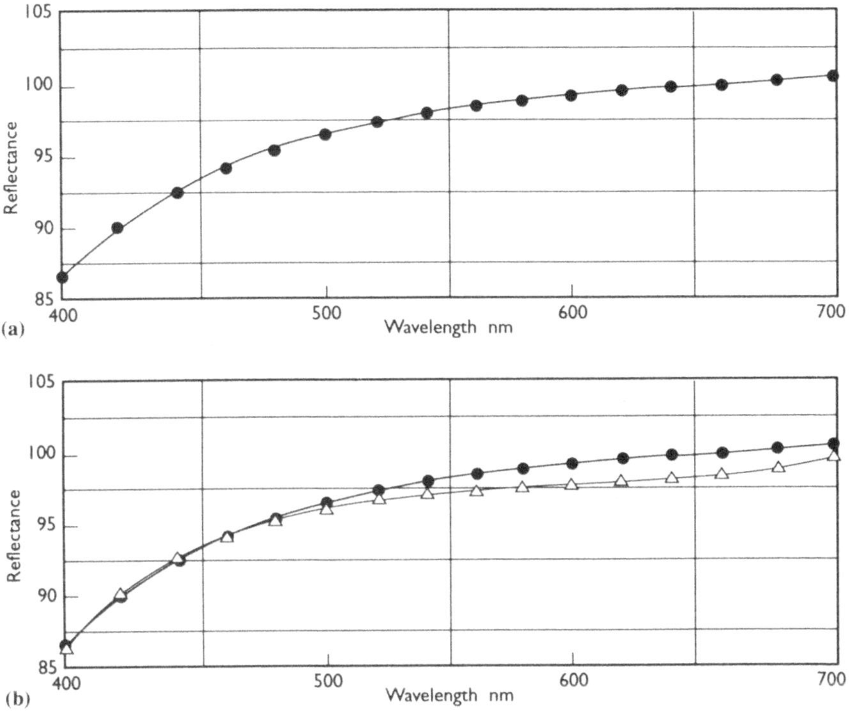

Figure 8.1 Reflectance curve of (a) virgin bleached pulp; and (b) pulp with (△) and without (●) 0.1% ultramarine dispersion.

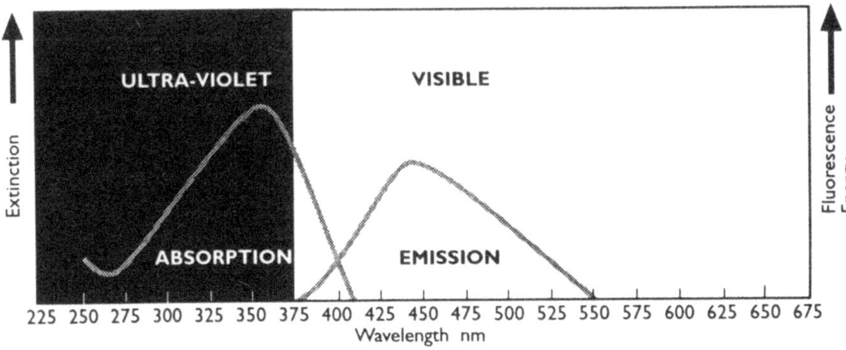

Figure 8.2 Ultraviolet absorption and visible fluorescent emission spectrum of fluorescent brightening agents.

ultraviolet content. Tungsten light is deficient in it, as is warm fluorescent illumination. On the other hand, artificial daylight illumination contains enough ultraviolet to activate FBAs effectively. The spectra of artificial illuminants are shown in Figure 8.4.

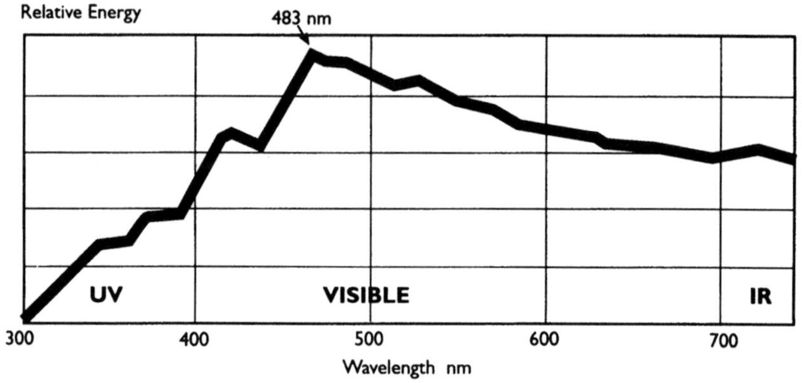

Figure 8.3 Spectrum of sunlight: 5% ultraviolet (UV); 40% visible; and 55% infra-red (IR) light.

8.3 The quantification of the effectiveness of fluorescent brightening agents

8.3.1 Quantum efficiency

The quantum efficiency of an FBA is a measure of the efficiency of conversion of ultraviolet light into blue; it is defined as the ratio of the number of blue light quanta emitted by fluorescence to the number of absorbed ultraviolet quanta. Effective commercial FBAs have high conversion efficiencies and quantum efficiencies of around unity [1].

8.3.2 Shade build curves

An important means of depicting the effectiveness of a brightener in use is through the 'shade build curve', which indicates the development of fluorescence intensity at increasing brightener additions. These curves normally show a maximum intensity, after which aggregation reduces the effectiveness.

8.4 The chemical nature of fluorescent brightening agents

FBAs are conjugated organic chemicals from one of several classes, for example, sulphonated stilbene triazines, benzoxazoles, coumarins and pyrazoline derivatives [1]. The class which dominates all cellulosic applications is that of the sulphonated stilbene triazines, as shown in Figure 8.5.

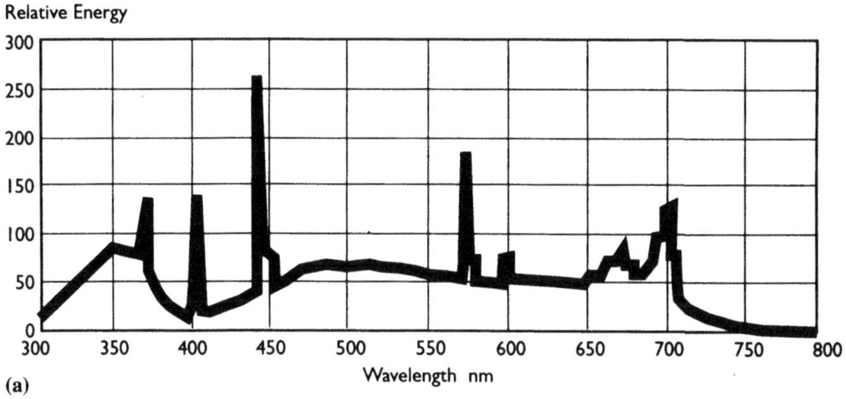

(a)

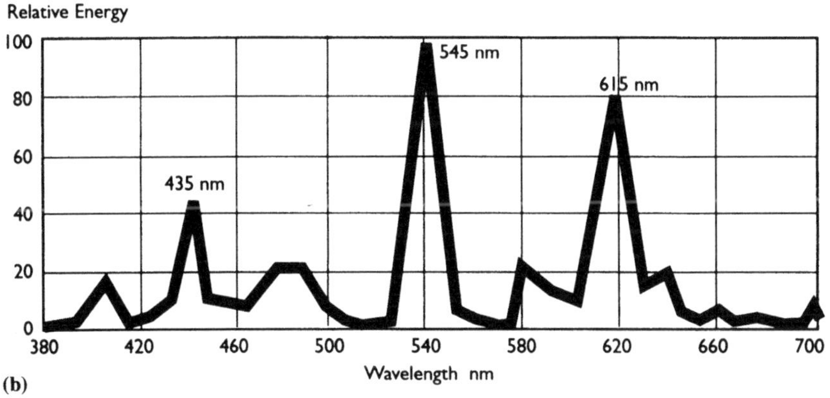

(b)

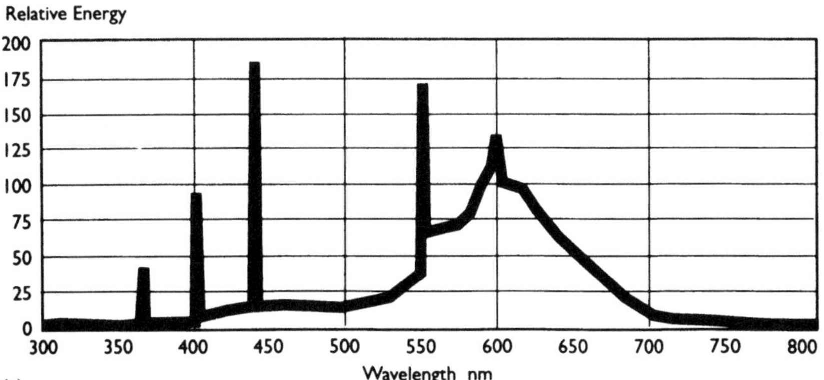

(c)

Figure 8.4 Spectral power distribution curves: (a) 'daylight' tube; (b) sales display lighting; and (c) warm white fluorescent lighting.

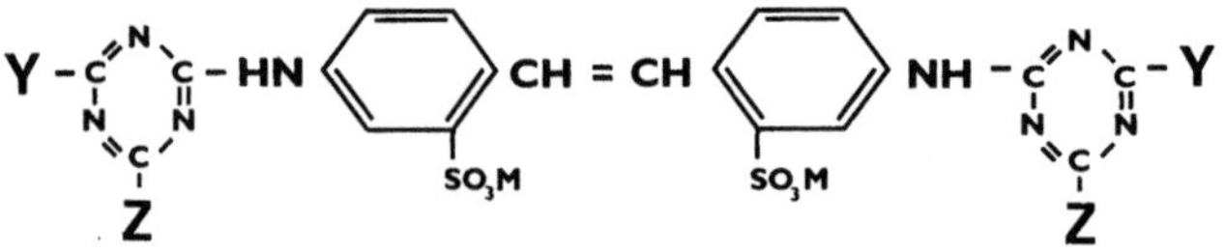

Figure 8.5 The generic structure of sulphonated stilbene triazines. M = for example, metal atom, amine group; for Y and Z in commercially important brighteners, see Table 8.1.

Table 8.1 Y and Z groups in commercially important sulphonated stilbene triazines, as illustrated in Figure 8.5

	disulpho	tetrasulpho	hexasupho
Y =	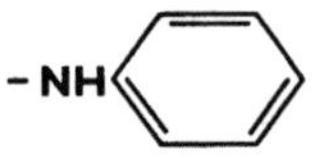	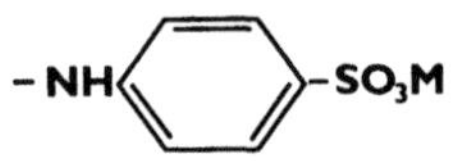	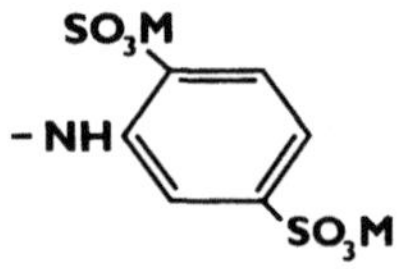
Z =	$-N(C_2H_4OH)(C_2H_4OH)$	$-N(C_2H_4OH)(C_2H_4OH)$	$-N(C_2H_4OH)(C_2H_4OH)$

Sulphonated stilbene triazines are water soluble and have an affinity for cellulose and polysaccharides, whereas the other classes have a specific affinity for various synthetic textile fibres.

They are synthesized from paranitrotoluene by multistage processes, including: sulphonation: oxidation: hydrogenation; and condensation with cyanuric chloride. Further condensations are made with primary and secondary amines.

Brighteners of this class are used also in textiles and detergents. Commercially important types for paper are described as disulpho, tetrasulpho and hexasulpho, indicating the number of sulphonic acid groups per molecule (Table 8.1). The number of sulphonic acid groups on an FBA determines its solubility in water, its cellulose substantivity and its effectiveness. Solubility, for its part, determines the tendency for a brightener to aggregate.

8.4.1 Disulpho brighteners

Disulpho brighteners are the least soluble in water of the three classes and the least effective weight for weight. They are also the most substantive to cellulose fibre, but the most prone to 'greening' (Section 8.6). This renders them really only suitable for addition to the papermaking stock. Surface application of these brighteners results in greening at very low addition levels, as shown in Figure 8.6.

These brighteners are more compatible with cationic additives than are either the tetrasulpho class or the hexasulpho class. Stock addition of disulpho brighteners can be used in situations where cationic quenching (reduced fluorescence because of interference by cationic additives: Section 8.6) is observed with the other classes. In such situations, it is important to use separate addition points for the brightener and the cationic additive.

8.4.2 Tetrasulpho brighteners

Tetrasulpho brighteners can be applied either to the stock or to the paper surface. They are less substantive to fibre than are the disulpho types, and more water soluble. It is thus easier to disperse them into the monomolecular layers required for effective fluorescence.

Because they are versatile, tetrasulpho brighteners are made in large tonnages and benefit from scale of manufacture, so they are relatively less expensive. As they are easily dispersed, they are economical in use. Grades are sold which are suitable for a wide variety of applications.

It has been established practice to add urea to the liquid formulations of tetrasulpho brighteners to aid product storage and low-temperature stability. This allows solutions to be stable at higher solids contents, though it does not enhance brightener performance.

Grades of tetrasulpho brighteners are now available which do not

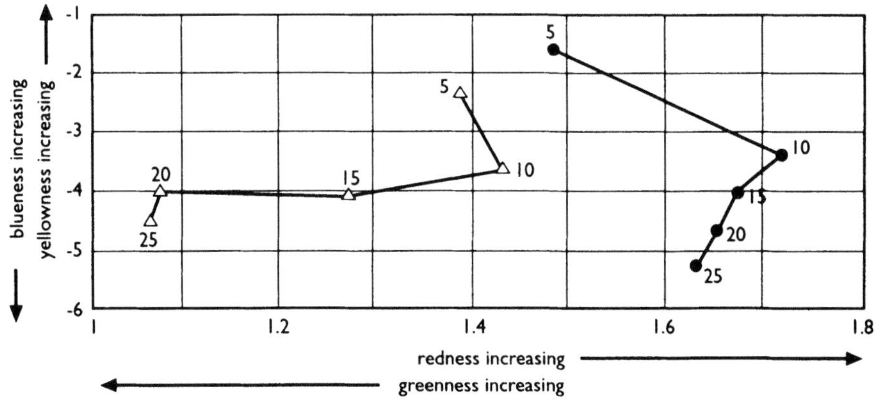

Figure 8.6 Size-press brightener application: disulpho (△) compared with tetrasulpho (●) brightener. Brightener concentrations (g l^{-1}) are shown by each point.

contain urea whilst retaining excellent solution stability. These formulations have the benefit of reduced nitrogenous load in paper mill effluent.

8.4.3 *Hexasulpho brighteners*

Hexasulpho brighteners are very soluble in water, have good resistance to acid and will build to give very high whites without greening. It is essential to apply hexasulpho brighteners to the surface as they lack sufficient substantivity to fibre for stock addition.

Hexasulpho brighteners are less economical than are the other classes of brighteners, being twice as expensive to produce as tetrasulpho brightener (per unit of active content). The raw material aniline 2,5-disulphonic acid is particularly expensive.

Typical shade build curves of tetrasulpho and hexasulpho brighteners are compared in Figure 8.7.

8.4.4 *Buffered brighteners*

Formulations of brighteners are available for surface application that include buffers to counteract acidity arising from cooking starch at the papermachine. The amount of buffer must be carefully controlled since it can have a serious adverse effect on the internal sizing of paper.

A comparison of the performance of buffered and unbuffered tetrasulpho brighteners after addition at an acid size press is shown in Figure 8.8.

Hexasulpho brighteners have greater acid stability and would only need to be buffered for the most extreme applications. Buffered grades of these brighteners are not commonly sold.

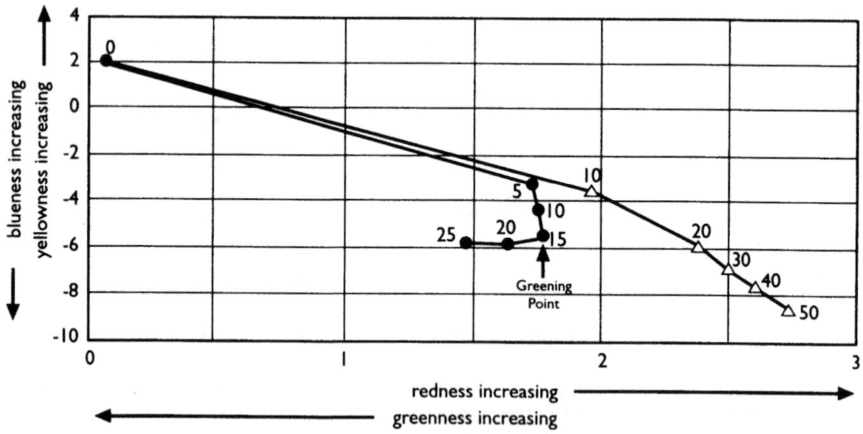

Figure 8.7 Brightener application: tetrasulpho (●) compared with hexasulpho (△) brightener, with starch size at 45°C. Brightener concentrations (g l^{-1}) are shown by each point.

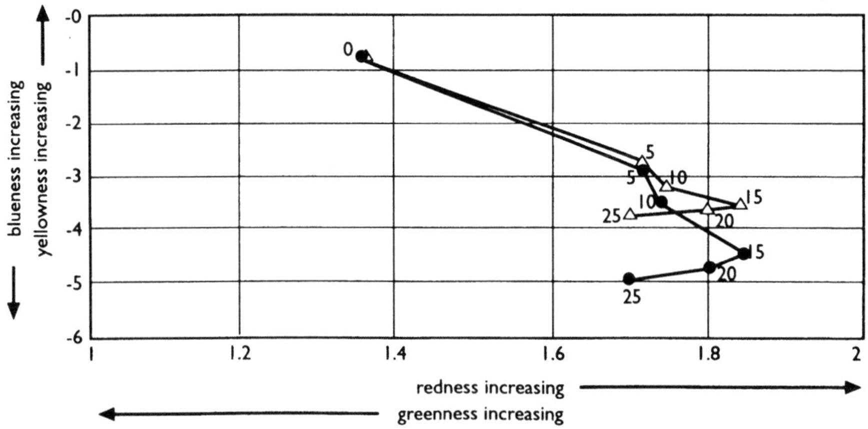

Figure 8.8 Brightener performance: tetrasulpho (△) compared with buffered tetrasulpho (●) brightener in the acid size press. Brightener concentrations (g l^{-1}) are shown by each point.

8.4.5 *Tetrasulpho/hexasulpho mixtures*

Although not widely used, such grades are available and offer an economic alternative for high white papers, as shown in Figure 8.9.

8.4.6 *Other brightener types*

Other types of high build stilbene triazine brightener are known but not yet produced commercially: for example, sterically enhanced tetrasulpho brighteners [2]. Others have been sold commercially but cannot meet the highest performance needs: for example, symmetrical hexasulpho brighteners.

Another type which is suitable for pigmented latex coating is 4,4′-bis(2-sulphostyryl)biphenyl [3]. Amino benzoic acid derivatives have also been used for brightening the surface of paper [4].

8.5 Application of fluorescent brightening agents to paper

As noted earlier, FBAs may be applied to paper either in the stock or at the surface. Application to the surface is advantageous since it ensures virtually 100% retention, but it is generally accepted to be necessary to add a proportion of the brightener to the stock. This ensures that a solid white is achieved and masks any unevenness caused by imperfections in the surface application. The normal split is around 4:1 in favour of surface application.

Control of whiteness and shade are both easier with surface addition,

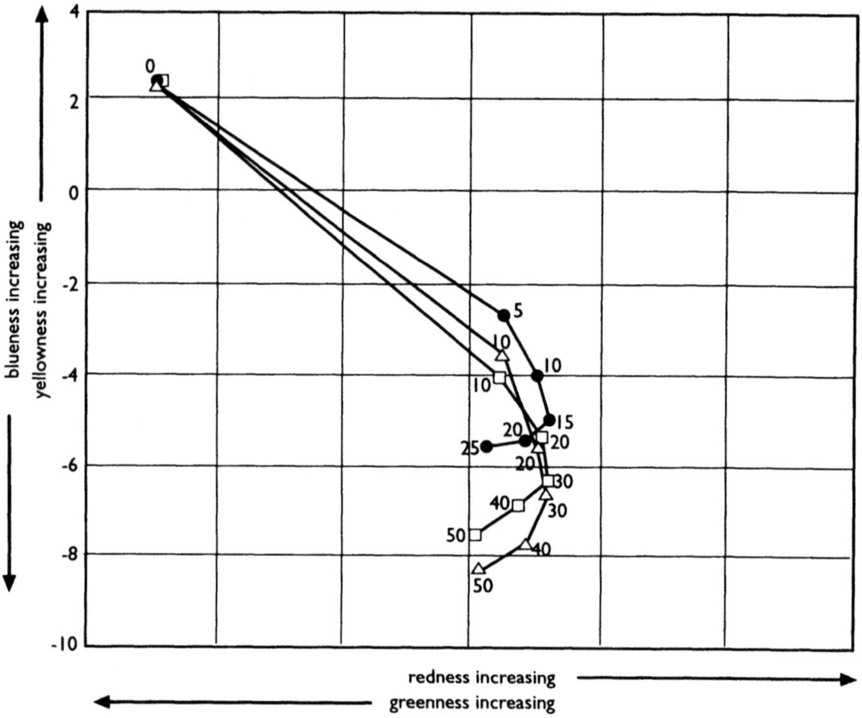

Figure 8.9 Brightener performance: comparison of tetrasulpho (●), hexasulpho (△) and tetrasulpho/hexasulpho mixture (□) brighteners in a high build size press. Brightener concentrations (g l^{-1}) are shown by each point.

though in papermachines with well-closed backwater systems, where there are minimal losses, addition to the stock can be almost as effective as to the surface.

Dispersion into monomolecular layers on fibre, important in ensuring effectiveness, is easier to achieve when the brightener is added to the stock.

8.5.1 *Surface application*

Surface application is made by using a carrier chemical: either on-machine, at the size press, film press, or blade coater; or in an off-machine coating operation. Carrier chemicals are usually starches or pigmented latex coatings, though carboxymethylcellulose (CMC) sodium salt or polyvinyl alcohol (PV alcohol) may also be used.

PV alcohol is the preferred brightener carrier in pigmented latex coating. It is rarely used in the size press and does not enhance brightener performance as would be expected by analogy with pigmented latex coating. If PV alcohol is used it is more appropriate to use a tetrasulpho brightener rather than a hexasulpho brightener. Figure 8.10 shows that a redder tone of white is observed with PV alcohol than with starch.

CMC is not commonly used at the size press but is a particularly effective brightener carrier. It can be seen from Figure 8.11 that CMC is particularly useful in high build applications. CMC elevates the greening point of both tetrasulpho brightener and hexasulpho brightener. Thus CMC may be considered for applications where the highest whiteness values are required. The benefits of using CMC in this context must, however, be balanced with economic, strength and coating mix viscosity requirements.

8.5.2 Film-press versus size-press application

The film-forming ability of the film press is very much superior to that of a size press, even when applying only simple starch. For this reason, film-press application is likely to give superior brightener performance to size-press treatment.

In addition, CMC is quite likely to be used when a film press is used as a pigmentizer, leading to further enhancement of brightener effectiveness.

8.5.3 Brightener addition levels

It has already been noted that it isn't possible to apply disulpho brighteners at the paper surface.

Recommended addition levels of concentrated tetrasulpho brighteners are in the range of 1–15 g l^{-1} starch. However, it is safe to apply most

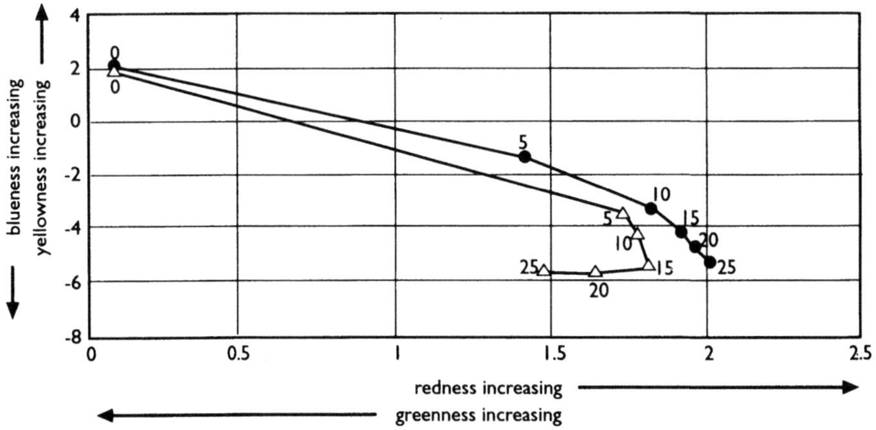

Figure 8.10 Tetrasulpho brightener application – a comparison of carrier chemicals: starch (△) compared with polyvinyl alcohol (●), at 45°C. Brightener concentrations (g l^{-1}) are shown by each point.

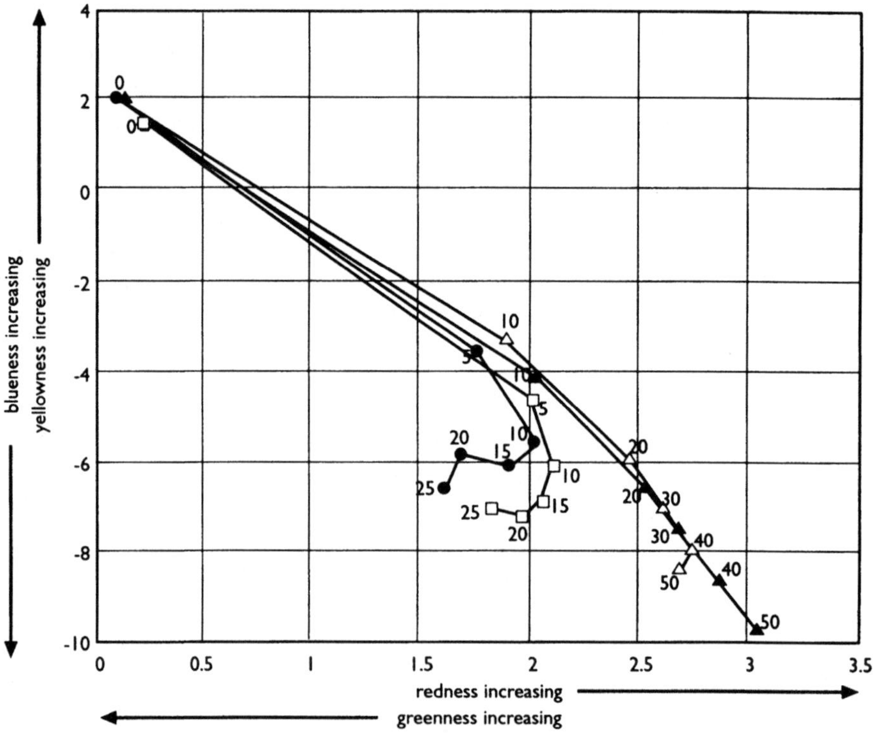

Figure 8.11 Brightener performance: comparison of tetrasulpho/starch mixture (●), hexasulpho/starch mixture (△), tetrasulpho/CMC mixture (□) and hexasulpho/CMC mixture (▲). Starch content = 9% at 60°C; carboxymethylcellulose (CMC) content = 3% at 60°C. Brightener concentrations (g l^{-1}) are shown by each point.

liquid formulations of hexasulpho brighteners at levels of up to 50 g l^{-1} starch. Careful assessment should be made to ensure that there is no adverse effect on the paper's internal sizing at high addition levels, since surfactant is used in the manufacture of some grades.

8.6 Impairment of fluorescent brightening agents: greening and quenching

There are various phenomena which impair the performance of FBAs. 'Greening' is particular to these chemicals and is the result of aggregation.

To function well, FBAs should form a monomolecular layer. If the brightener molecules aggregate, for whatever reason, then the wavelength of the re-emitted light becomes longer, making the fluorescence shade greener and less blue.

Greening can be induced by any of the following factors, which tend to affect brighteners added at the surface rather more than those added to the stock:

- high concentration of FBA;
- acidity;
- low starch temperature;
- cationicity.

8.6.1 Concentration of fluorescent brightening agents

Brighteners need to be distributed in as close to monomolecular layers as possible for optimum effectiveness. The reason that shade build curves exhibit a maximum is probably because the brightener molecules begin to cluster as their concentration increases.

Increased water solubility helps to achieve good brightener distribution, though too much can mean excessive drain losses. Because it is much easier to achieve an even, monomolecular distribution from an aqueous solution, this problem is easier to avoid with addition to the stock rather than to the surface.

8.6.2 Acidity

It has already been noted that acidity (sometimes resulting from starch cooking at the papermachine) can severely impair brightener effectiveness. Indeed, buffered formulations of tetrasulpho brighteners are available to overcome this, which is a problem mainly affecting surface addition.

8.6.3 Starch temperature

Another problem that can affect surface addition of brighteners is low starch temperature in the size-press or film-press mixture.

Greening can be reduced by running the size press as hot as possible. Figure 8.12 demonstrates that at 60°C greening is less than at 45°C.

8.6.4 Cationicity

Cationic agents can severely affect the performance of tetrasulpho and hexasulpho brighteners. Even the less anionic disulpho brighteners are adversely affected by highly cationic systems.

Surface addition does tend to reduce the incompatibilities between brighteners and cationic stock additives; though some wet-end additives, such as sizing cure promoters, are still active enough to impair brightener efficiency even with surface application. Certainly, the use of cationic starch at the size press strongly interferes with brightener performance.

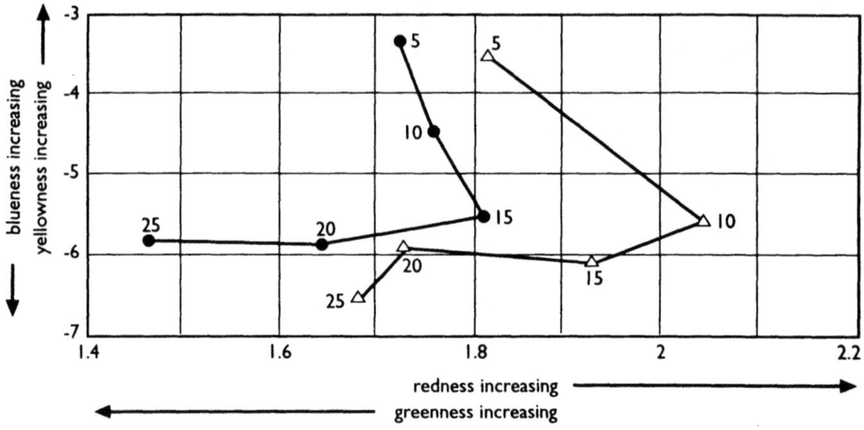

Figure 8.12 Tetrasulpho brightener application at starch temperatures of 45°C (●) and 60°C (△). Brightener concentrations (g l^{-1}) are shown by each point.

8.6.5 Other interfering substances

Apart from greening, brightener performance can be impaired by the presence of transition metal ions, by competition from other ultraviolet absorbers and by strong oxidants (which render the molecule inactive).

Hard water can affect the performance of disulpho brighteners, though this is rarely a problem with tetrasulpho and hexasulpho brighteners, whose calcium and magnesium salts remain adequately soluble. Thus, this problem does not normally affect surface added brighteners.

Heavy metals, especially copper, can severely reduce the performance of tetrasulpho and hexasulpho brighteners. It has happened that copper ions from brass fittings on the starch cooker have had a noticeable impact on surface added brighteners.

Examples of other ultraviolet absorbers in competition with the brighteners are

- lignin (brighteners do not function well in papers containing mechanical pulp);
- both forms of titanium dioxide (anatase less than rutile);
- some of the less brightly coloured mineral fillers;
- process-water containing dissolved and suspended coloured substances, such as peat (this problem tends not to be so important to surface added brighteners).

Finally, free hypochlorite attacks aqueous FBAs, severely reducing performance.

8.7 Lightfastness

8.7.1 Lightfastness of brighteners in solution

Dilute solutions of stilbene triazine brighteners are unstable to irradiation by ultraviolet light. This is because of an isomeric transformation. The normal and useful isomer form of the brightener molecule has the *trans* structure (Figure 8.13(a)) in which the stilbene linkage takes a chair-like configuration. After ultraviolet irradiation the stilbene linkage takes on the boat-like *cis* configuration (Figure 8.13(b)). Unfortunately, this isomer lacks substantivity to fibre and does not absorb ultraviolet light at useful wavelengths. Thus, laboratory evaluations of brighteners must be carried out with great care. Dilute solutions must be prepared and stored in ultraviolet-free environments. Wherever possible, it is better to use undiluted products to eliminate isomerism errors. Fortunately, the lightfastness of FBAs is quite satisfactory after addition to starch, PV alcohol and CMC solutions. Thus laboratory simulations of surface application can be carried out. Fortunately, too, lightfastness of dilute solutions is a problem not commonly encountered in full-scale paper-making.

8.7.2 Lightfastness of brightened paper

Lightfastness of brightened paper is an important quality consideration. It is normally assessed by exposure to artificial daylight according to ISO Method 105 BO2, which has a scale of 1–8 (1 is bad, 8 is excellent). The method was designed for coloured papers and is not actually very easy to carry out reproducibly on brightened papers.

The result depends on the lightfastness of the substrate, surface additives and FBA. Results of 3 or 4 on the blue wool scale are considered acceptable and are usual for wood-free brightened papers.

Newsprints typically have lightfastness values of less than one. Unbrightened acrylic coated wood-free paper, however, can have a lightfastness of 5; and polyester cloth, brightened with benzoxazole brightener, can have a lightfastness of 7.

Lightfastness may be somewhat reduced by surface application of the brighteners, but the effect is not marked.

Figure 8.13 (a) *Trans* and (b) *cis* configurations of the stilbene linkage.

8.8 Recycling

Large tonnages of brightened papers are recycled, both as 'own-broke' and as bought-in waste paper. Such papers may have been brightened both at the surface and in the stock.

8.8.1 Charge considerations

Recycled stock is weakly charged either positively or negatively, according to the balance of the charges on the fibres, the starch, the brightener, and on any other components present. FBA has a weak negative charge. The stock may thus have a negative charge if a lot of brightener has been used. The charge from the brightener will be enhanced in stock which is warm and alkaline, since such conditions encourage ionization of the sulphonic acid groups, which also improves solubility. As might be expected, hexasulpho brighteners impart more charge to stock than do tetrasulpho brighteners, while the disulpho class is the least likely to impart any charge.

In order to deal with problems arising from excessive negative charge, polymer suppliers provide proprietary 'anionic-trash catching chemicals' to cancel out the charge. Unfortunately, such anionic-trash catchers are cationically charged, so that when used in commercially useful quantities they act like other cationic additives, impairing brightening performance.

Provided cationic polymers are not used, there is no reason why the brightening effect should be diminished by recycling. Indeed, recycling may actually encourage brightener dispersion and lead to improved efficiency.

8.8.2 Residual fluorescence

Residual brightener effect can be a drawback in certain situations, for example, when using recycled fibre for the production of coloured papers. In this situation, non-fluorescent stock may be required and it becomes necessary to use a quenching agent to kill the fluorescence. Such agents are formulated polymeric imidazoline derivates which interact with FBAs to reduce or completely eliminate brightening.

Quenching efficiency depends on the type of brightening agent present and on the amount of quenching agent used. Tetrasulpho and hexasulpho brighteners are easier to quench than are disulpho brighteners. It may be necessary to use up to 3% quenching agent to mask completely the fluorescence of very bright paper made with disulpho brightener.

Other applications, for example, medical papers, may require totally brightener-free stock, which has led to attempts to find methods of chemically destroying FBAs *in situ*. Unfortunately, no simple, practical methods are available.

The only methods which work utilize strong oxidizing agents such as sodium hypochlorite or potassium permanganate [5]. The potassium permanganate method is the more effective method but is messy and impractical. Sodium hypochlorite is easier to use but acts only on brightener in the aqueous phase. Thus it is effective against tetrasulpho and hexasulpho brighteners in warm alkaline stock, but less so against disulpho brightener. The effluent produced by this process has not been characterized.

Surface applied hexasulpho brighteners are probably less susceptible to the problem of residual fluorescence because of their greater solubility in water. Pulping conditions for many recycled furnishes will greatly favour the dissolution of FBAs, enabling their removal with the water during any washing stages in fibre-recovery processing.

8.9 Monitoring and control of surface brightened paper

Before paper brightness can be controlled successfully it must be accurately measured, and there are several important considerations involved in this. The following comments refer mainly to laboratory-based instruments, but are also true of on-machine gauges.

8.9.1 Instrument specification

The specification of the instrument used is the first consideration. It should satisfy the following requirements:

- use appropriate illumination;
- be stable;
- use appropriate geometry;
- be easily calibrated.

8.9.2 Appropriate illumination

It almost goes without saying that it is important that the illuminant used should be rich in ultraviolet, in the wavelength range from 350–400 nm; it is preferable that the illuminant should conform to the specification for simulated daylight D65.

Xenon flash illuminants are now the most common, though calibration of the ultraviolet output of these can be problematical since it is difficult to know when the lamp has started to deteriorate.

Continuous xenon illumination has been used extensively: the drawback with this is that the lamp gets very hot and is not robust.

A less sophisticated and less stable illuminant is a standard artificial daylight fluorescent tube.

Tungsten and tungsten halide illuminants are yellower and deficient in ultraviolet, and are hence unsuitable for monitoring FBAs.

8.9.3 *Stability*

Apart from illuminant stability (considered above) the rest of the instrument should also be stable enough to withstand paper mill conditions. It must be especially tolerant of supply voltage fluctuation.

8.9.4 *Geometry*

Ideally, instruments should include integrating spheres in their optics in order to reduce the problems caused by sample orientation. Integrating spheres make it possible to present samples for measurement in either the machine or cross-direction, with similar results.

8.9.5 *Calibration*

Standard white fluorescent cloths are available for calibration. Proprietary white fluorescent plastic tiles can also be used. Barium sulphate is the normal standard for non-fluorescent whites, but it cannot be used for calibration of the ultraviolet output of lamps.

8.9.6 *Cleanliness*

Dust can be problematical. It is important to keep instruments in a clean environment, preferably in controlled-condition laboratories.

8.9.7 *Measurement*

Various measurements have been used over the years to assess the performance of FBAs. The simplest are reflectance at 457 nm (brightness) and Z reflectance. These give indications of the increase in reflectance in the blue visible part of the spectrum. Higher values indicate an FBA effect.

Direct measurement of fluorescence is another simple measurement. In this, the paper is illuminated with pure near ultraviolet light. FBA on the paper is activated and the reflected blue white light is detected by a photocell calibrated in Harrison fluorescence units. A simple measurement of the FBA effect is thus obtained.

However, different papers, with similar blue reflectance and fluorescence values can have very different appearances. Whiteness indices, such as CIE (Commission International d'Eclairage) Whiteness [6], help to overcome this deficiency because they describe the whole colour of the paper. There are various different whiteness indices and they have different hue biases. CIE whiteness is the most commonly used. In this, the lightness and shade of paper are described by *L-a-b* measurements. Lightness (*L*) is only marginally changed by FBA effect. Shade, however, is very sensitive to FBA. A plot of redness (*a*) and blueness (*b*) can be used to give a sensitive measurement of brightener performance. However, shade measurements are only reliable when made on instruments whose lamps are in good condition and whose ultraviolet output is calibrated. Measurements of whiteness indices are not sensitive to small changes in FBA effect.

8.9.8 Control

The use of modern on-machine instrumentation and continuous dosing equipment has made continuous monitoring with computer control of paper brightness possible. It is usual to implement this with split addition, with some brightener added to the stock and the rest to the surface. The split ratio is often around 4:1 in favour of the surface. Maintaining the addition rate to the stock constant while controlling the addition to the surface gives quick response with good control.

However, whilst continuous addition is generally beneficial, there is a real danger of overdosing. Because duller papers are perceived by the control system as having greater FBA demand, corrective overdosing can occur. Overdosing with FBA actually makes paper duller still. Continuous addition systems must be controlled so that manual override is needed above a pre-set addition limit. The pre-set limit should be determined by mill conditions and target brightness.

8.10 Toxicology and ecotoxicity of fluorescent brightening agents

Stilbene triazine FBAs for detergent applications have been in commercial production since the early 1950s. Latterly the same and similar brightening agents have been introduced to the paper industry. No known adverse effects have been encountered by production workers in the producing factories.

FBAs have been extremely widely studied, leading to a wealth of published information [7]. They are considered to be innocuous chemicals which harm neither humans nor the environment. Indeed, FBAs have been recommended as water tracing chemicals [8].

8.11 Summary of the effects and influences on surface addition of fluorescent brightening agents to paper

In summary, then, brightening of paper by surface application of FBAs is characterized by the following:

- the most effective FBAs are only usable in surface application;
- very high addition rates of these FBAs can be used;

- the use of CMC as carrier further enhances their performance;
- the retention of chemicals applied this way is virtually 100%;
- the use of film-press technology will further enhance the FBA effectiveness by concentrating it at the surface.

However, the following conditions need to be met to ensure successful surface application:

- size or film-press mixture temperature should be high (preferably above 50°C);
- the use of cationic surface additives as well needs to be very carefully monitored. It will probably not be possible for very high brightening;
- starch preparation should be carefully controlled in order to avoid acidification;
- care needs to be exercised to minimize the possibilities of contamination by transition or heavy metals;
- very high addition rates of FBA, even at the surface, can degrade paper sizing.

Acknowledgements

Thanks are given to Mrs Linda Adams for her comprehensive size-press application studies. The assistance of Mr Paul Cutts of Allied Colloids in providing valuable insights into the charge effects demonstrated by brightened broke is greatly appreciated.

References

1. Anliker, R. and Müller, G. (1975) *Fluorescent Whitening Agents*, Georg Thieme, Stuttgart.
2. Thompson, J.D. (1995) Sulphonated triazinyl stilbene derivatives and their uses in optical whitening agents. GB patent application 95/01428.
3. Traugott, W., Rohringer, P., Traser, G. and von Raven, A. (1994) Einsatz neuer Aufhellersysteme bei gestreichenen Papieren. *Wochenblatt für Papierfabrikation,* **6**, 224–8.
4. Gowan, J.W. Jr and Lo, D.K. Fluorescent whitening agents: products comprising the agents and methods of use thereof. US patent 5 026 507.
5. Williamson, R. (1980) *Textile Science and Technology: Fluorescent Brightening Agents*, Elsevier, Amsterdam.
6. McLaren, K. (1983) *The Colour Science of Dyes and Pigments*, Adam Hilger, Bristol.
7. Kramer, J.B. (1992) Fluorescent whitening agents, in *The Handbook of Environmental Chemistry*, vol. 3, part F (eds O. Hutzinger and N.T. de Oude), University of Amsterdam Press, Amsterdam, pp. 351–66.
8. Aldous, P.J., Fowell, J.K. and Hunt, S.M. (1987) *The application and toxicity of Photine C when used as a groundwater tracer*. WRc Environment, Medmenham, UK, June.

9 On-machine surface coloration

A.S. TINDAL

9.1 Introduction

Commercially produced paper can be dyed using three methods:

- stock dyeing, where the dye is applied to the paper fibres throughout the sheet;
- surface dyeing, where the dye is applied at the surface only;
- split addition, a combination of the two.

The primary method of colouring paper is stock dyeing, where advances in dye chemistry and formulation together with continuous dosing equipment and closed-loop colour control technology have resulted in systems which can even produce deep shades without backwater coloration and with the minimum of wastage. In practice, paper coloration at the surface is not widespread when compared with stock dyeing, but it has proven successful and can provide opportunities for the papermaker, providing the potential advantages and disadvantges have been carefully evaluated.

There are several ways in which dyes and pigments can be applied to the surface of the paper, such as size-press, spray dyeing, calender stack dyeing and coating slip dyeing. The scope of this chapter will be limited to on-machine surface dyeing with particular emphasis being given to applications at the size press and in paper coatings. The more speciality applications for the size press and coatings will not be covered such as carbonless colour-formers; on-machine printing and patterning processes; transfer papers; heat and pressure sensitive papers; security products. Addition of dye to the coating slip, on-machine, either as a shading dye for high whiteness papers or directly to produce a particular shade, is also an important commercial field.

The coloured materials most commonly supplied to the industry are liquid formulations of water-soluble dyes, although powders and granulated dyes are used. Coloured pigment powders and suspensions are also used but the term pigment, used here to describe an insoluble synthetic coloured dispersion or powder, should not be confused with other uses for the word in the paper industry such as coating pigments (calcium carbonate, talc, clay or titanium dioxide).

9.2 Size-press dyeing

In the 1970s many references appeared worldwide highlighting the potential advantages of size-press dyeing [1–11], but this approach still only accounts for a relatively small volume of coloured paper production. The main reasons for size-press dyeing not achieving the success predicted are:

- stock dyeing technology overhauled many of the advantages offered by size-press dyeing;
- the achievement of uniformity and levelness of dyeings was a much greater problem than at first envisaged;
- colour control and colour changes were more difficult than when using continuous stock dyeing;
- bleed-fastness properties of size-press dyed papers were invariably inferior.

Size-press dyeing therefore remains an important alternative to stock dyeing rather than the dominant method of paper coloration.

In size-press coloration, dyes or pigments are added to the sizing solution which is then transferred to the surface of the paper using either a conventional size press or a film press. The main feature of this type of dyeing is that the substrate to which the dye is initially applied (in the case of direct dyes) is the starch itself. The direct dye bonds to the starch in much the same way as dyes bond to cellulose fibres when added at the wet end [12]. The coloured starch solution is absorbed into the paper as it passes through the size-press nip and the paper is then dried. Clearly, the many factors which affect starch absorption into the paper sheet determine the extent of the dyeing in terms of penetration into the paper, levelness and depth of dyeing. Generally, the more the starch solution is uniformly absorbed the better the quality of the dyeing.

Once this general process is understood, the difficulties involved with size-press dyeing can be predicted: the previously clear starch sizing layer is now stained with colour so that it can be clearly seen, as of course can any imperfections in the uniformity of the coating. These imperfections, which may have been previously invisible, are magnified substantially by the contrast with the base sheet which is usually white. In addition, any factors which change the starch pick-up through changing the absorbency of the base sheet will change the dyed appearance of the sheet. As the starch is strongly coloured, even small changes can be detected by the eye. The influencing factors include:

- the amount and evenness of the internal sizing;
- the amount and type of broke used;
- the furnish composition;
- formation;
- other factors which are frequently variable during a typical colour run.

The sizing solution in the size-press supply system is also coloured and must be dealt with before a colour change is made. It should also be noted that, although the starch solution usually contains a relatively large amount of dye, it remains transparent, therefore it cannot be used to hide surface imperfections such as specks and dark fibres in the base sheet completely.

9.2.1 Selection of dyes and pigments

In the selection of suitable dyes and pigments for use at the size press care must be taken to ensure that they are compatible with the starch and any other additives that are being used. It is advisable to check compatibility on a limited scale as precipitation of the sizing solution can occur in some cases. There are several classes of dye available, the chemistry and properties of which have been discussed previously [12]. It should be noted that the colours achieved by size-press dyeing may be different to those shown in the shade cards of the respective dye suppliers since these apply to stock dyeing. Since the dye is now applied to the starch at the surface, the bleedfastness will also become a function of how easily the starch, with the dye attached, can be removed. In addition, other factors, such as lightfastness, may also be affected by surface application. It is important to note that some of the dye classes are not compatible in solution, and for this reason dyes of different classes should not normally be used together in the size press.

Anionic direct dyes form the main class of dyes used at the size press today. They have good substantivity for most starches and are compatible with most starch systems except those which are strongly cationic. They are available in a wide range of colours and can be selected according to the requirements of the finished paper for bleedfastness, substantivity, lightfastness, bleachability, etc.

Cationic versions of direct dyes are available, though the range of colours can be quite small in some cases. They have excellent substantivity for most starch types. Compatibility with the size-press solutions is more limited than with the anionic direct dyes, however; but they can also be used with most cationic starch systems. Not all cationic direct dyes are compatible in solution with each other, therefore recipes must be thoroughly checked for stability in the laboratory before using on-machine.

Cationic basic dyes are inexpensive to use and have a brilliant shade range. They are less technically advanced than the anionic and cationic direct dyes and difficult to produce in a liquid form. For this reason, basic dyes available in liquid form often contain large amounts of acid or other solubilizing systems which may affect the sizing solution, in particular by lowering the pH. The lightfastness of these dyes is very poor and their use is therefore limited to applications where this particular property is not required. For best results the furnish should contain more than 30% of a wood-containing pulp since they generally are less substantive to the starch and therefore are absorbed onto the paper fibres after passing through the size-press.

Pigments are insoluble, coloured materials which are supplied as aqueous dispersions. Unlike dyes, they are not soluble in the sizing solution. Lightfastness is usually high, as is the bleedfastness, providing the starch coating is well attached to the paper and cannot be easily removed under the particular conditions used in the fastness test. Pigments have no substantivity for the starch or fibres in the paper sheet and are therefore physically trapped at the surface purely by being immobilized by the starch layer. Pigments are not bleachable and are usually expensive. They are not commonly used in the size press although certain fluorescent pigments are available for use in, for example, security and stamp papers.

The solubility of acid dyes is very high and they can only be used in paper stock where fixing agents or alum are used as well. For this reason, and because of their very poor fastness properties they are not normally recommended for size-press use.

9.2.2 *Practicalities of size-press dyeing*

There are a number of key factors that need to be considered if size-press coloration is to be successful, and these are summarized in Table 9.1 [1, 13].

Dyes can be added to the size-press supply system at a suitable point by continuous addition. Liquid dye types can be dosed using a similar pumping system to that used in continuous stock dyeing [12]. The response to changes made at the dosing station is dependent on the total volume of the starch circulation system and the point at which the dye is added. These response times are usually greater than for a corresponding colour correction with a conventional wet-end continuous dyeing system. Colour changes, start-up times and clean-down times must all be considered in the total cost of size-press dyeings.

The uniformity and absorbency of the base paper plays a critical role in the effectiveness of size-press dyeing. Any factors which can change the absorbency of the base sheet must remain stable throughout the colour run for the paper to be of uniform appearance. A particular problem of size-press dyeing is that the results can vary from run to run. This is usually because of variation in the absorption of the base sheet from important factors listed in Table 9.1 [1]. Note that most of these factors simply affect the absorption of the starch solution at the size press [13].

Where a mixture of different fibres is used in the furnish, for example, in recycled paper containing some darker fibres, the natural colour of the fibres is not covered fully unless the colour applied is very deep. In

Table 9.1 Factors influencing the results obtained in size-press dyeing. Sources: Arnold, E. and Martin, G., Possibilities and limitations of size-press dyeing, *Wochenbl. Papierfabr.*, **122**, (9–10), 381–91; Clariant (Sandoz) Paper, Technical booklet 8B 5944/75, 1975

Factor	Disadvantageous	Advantageous	Notes
Base-sheet properties:			
Pulp type	Long fibre Sulphate pulp Groundwood	Short fibre Sulphite pulp	See Arnold and Martin (1994)
Ash content	None	10% or more	
Degree of beating	> 35 SR	< 35 SR	
Sized broke	> 10%	0%	
Formation	Non-uniform	Uniform	
Pick-up of size solution	< 25%	> 25%	Wet pick-up by weight of paper
Substance	> 100 gsm	< 100 gsm	Note, however, that dyeing costs are relatively cheaper than stock dyeing the heavier the paper
Other factors:			
Starch concentration	< 4%	10% or greater	
Starch temperature	< 40°C	> 40°C	Lowers the viscosity of the solution
Other additives		Synthetic binders such as polyvinyl alcohol and carboxymethyl cellulose (CMC)	See Section 9.2.3

addition, variation in the absorbency of the individual fibres may give a mottled appearance.

It is possible for dyeing at the size press to produce different colour strength results from one side of the paper to the other. This two-sidedness occurs when the absorbence of the top side of the sheet is sufficiently different from that of the wire side to produce a visible effect in the dyeing. It is possible to correct two-sidedness by dosing the dye separately to each side of the size press, provided the configuration of the size press and the starch system allow.

Best results are achieved when size-press dyeing is used together with continuous dye addition in the stock. This split-addition method results in greater colour control and a more uniform appearance since the contrast between the base paper and the surface colour is less noticeable.

A particular aspect of dye use which is not encountered in stock dyeing is that of bronzing. Bronzing can occur in deep shades where there is more dye present that can be adsorbed by the starch. In this case there are dye molecules present in a free state unattached to starch. These unattached molecules cannot be detected until drying occurs, at which point the dye in solution is precipitated onto the sheet as agglomerates of dye molecules rather than as individual dye molecules adsorbed onto the starch. These agglomerates form particles of solid dye during drying. Since the solid dye colour is not the brilliant shade seen when the dye is either in solution or absorbed onto a suitable substrate, this results in a dull, often metallic, cast, referred to as bronzing. This effect is dependent on the particular dye being used and the amount of starch present.

Some dye formulations produce foam when added in large amounts to the sizing solution. Any tendency of the dye to foam will be exaggerated compared with stock application. A suitable anti-foam, compatible with the dye and the size-press additives, must then be used. In certain cases foaming may not be controlled by the use of anti-foam, in these cases changing the dye or combination of dyes used may reduce the problem.

Once a colour run has been performed the problem of washing down the system occurs. Usually the washing down process is made faster with the use of a suitable bleaching agent, provided this does not represent a serious hazard and is known to be compatible with the other materials present, for example, in the size-press rolls and service system. The bleaching agents used are dependent upon local legislation; however, the three most common are hypochlorite, hydrosulphite (dithionite), and FAS (formamidine sulphinic acid). The supplier's instructions for the handling and use of bleaching chemicals should always be followed.

It should be noted that since many of the problems associated with size-press dyeing are due to the highlighting of non-uniformity in the starch coating, dyeings of high lightness show less contrast than those of low lightness. For this reason it is much easier to achieve apparently uniform results with bright yellow shades than with any other colour. The order of difficulty is as follows with the easiest listed first [13]:

$$\text{yellow} > \text{orange} > \text{red} > \text{blue}.$$

It is commonly noticed in practice that yellow and orange shades can be produced easily but that problems occur with red and blue shades (Plate 3).

In certain cases the dyed paper can have an 'orange peel' appearance (Plate 4). This means that the colour is not uniform and shows a regular pattern of small patches where the base sheet shows through, giving the appearance of orange peel. This pattern is caused by problems in the film-splitting of the starch solution in the outgoing nip of the size press. It is made worse by non-uniform formation. To decrease this effect the absorption of the coloured starch solution must be increased, or preferably some dye must be added to the stock by continuous dyeing to reduce the contrast between the base sheet and the surface colour (Plate 5). This effect is not normally visible, even when the starch contains FBAs (fluorescent brightening agents); however, the staining of the starch coating makes it visible to the eye.

9.2.3 Promoters for size-press dyeing

There are additives which can improve the quality of the dyeings produced. Some are more practically useful than others [1] and the addition of any new component to the starch system must be checked thoroughly for compatibility with the equipment and process as well as for the effect on other paper properties, aside from colour. Health and safety consequences must also be considered.

Film formers such as carboxymethyl cellulose (CMC) and polyvinyl alcohol (PV alcohol), promote the uniformity of dyeing in the size press by providing an alternative substrate to the starch which the dye can absorb onto. In addition they can provide a more uniform film, thereby increasing colour uniformity. These types of products can have a large effect on the viscosity of the size-press solution.

Wet-strength agents can be used in the starch solution to improve the bleedfastness properties of the dyeing since they prevent the starch coating from being easily removed by water [13].

Fixing agents can be used in certain circumstances to improve the bleedfastness of size-press dyeings still further; however, this necessitates a secondary process, as fixing agents cannot be used in the same starch solution as the dyes.

9.2.4 Advantages and disadvantages of size-press dyeing

Size-press dyeing is successful in the production of some papers, either because it is more economic or because it enables results which cannot be achieved when stock dyeing is used. The following advantages can be realized.

- Reduction in dyestuff usage: by applying the dye to the surface of the paper instead of the entire interior, savings can be made in dye usage. These savings are dependent on the grammage of the paper and the depth of shade produced (Figures 9.1 and 9.2) [13].
- Depth of shade: when producing particularly deep shades it is sometimes impossible to reach the depth required using continuous dyeing alone because of the relatively short contact time the dye has with the fibre and the capacity of the pumps available. In this case another method must be used as well. There are two available methods for producing deep dyeings. One is batchwise addition at the pulpers, thereby giving the dye more time to adsorb onto the fibres. The other is size-press dyeing, which can often produce deeper dyeings than stock dyeing alone, particularly when the shade required is close to or at the saturation limit of the dye when applied in the stock. In this case the base sheet is first coloured to a paler shade by means of continuous dyeing in the stock.

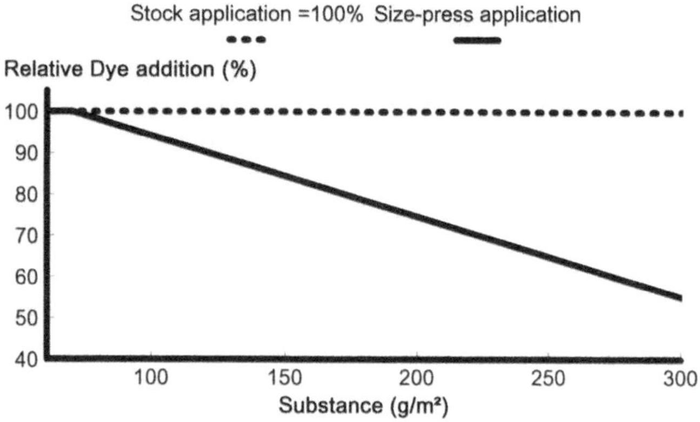

Figure 9.1 Relative dye usage to achieve a similar colour strength.

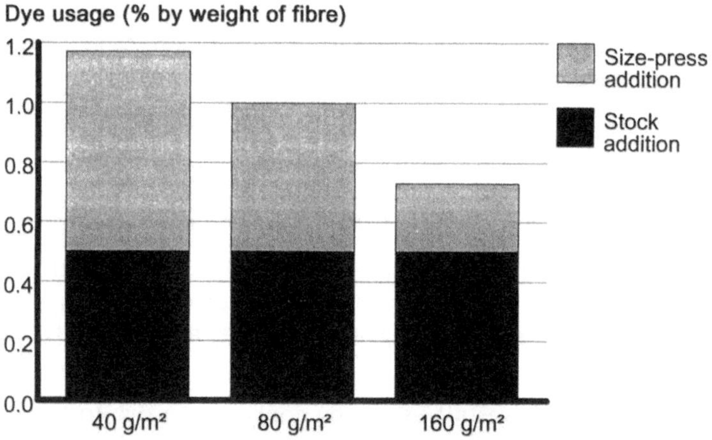

Figure 9.2 Relative dye usage to achieve a similar colour strength on three red papers of different substance dyed to the same colour strength in the stock.

- Special effects: some paper producers use dyes at the size press to produce ingenious special coloured papers. Since most of these methods are closely guarded secrets they are not discussed here; however, effects available are:
  - single-side coloured, where only one side of the paper is coloured;
  - deliberately non-uniform appearances, such as blotches or cloudy appearances;
  - security papers, in which fluorescent pigments enable detection in stamp papers, for example.
- Shading dyes for high white grades: although the preferred addition point for shading dyes is in the stock, they can also be used in the size

press. The effect is so subtle that the problems of uniformity are not encountered as they cannot be picked up by the eye because of the low concentration of colour.

- Correcting two-sidedness: size-press dyeing is used on occasion to correct coloured two-sidedness produced by dyeing in the stock. This can be achieved where the pick-up of the starch can be made different from the wire side to the top side or where the dye can be dosed to one side of the sheet only via the size press. Since the addition is relatively low and the contrast with the base sheet so small, the problems of uniformity are, again, not encountered. Usually, however, two-sidedness problems in the stock can be improved by changing the stock dyeing recipe by using a technique such as the Clariant pulsator method.

Many of the advantages predicted for size-press dyeing in the 1970s have been overtaken by advances in dye chemistry, formulation and continuous dyeing technique. In particular, modern dyeing techniques can now even produce deep shades without producing coloured backwaters. Therefore this list of advantages is smaller than those discussed in a typical paper published at that time [1].

There are also drawbacks to size-press dyeing, some of which are listed below.

- Uniformity of colour: problems with the uniformity of the starch coating, which were previously invisible, now become obvious to the naked eye. For this reason all equipment must be in good condition for best results and the surfaces of the size-press rollers must be relatively unblemished. Problems that can be highlighted by the coloration of the starch coating are:
  - unlevelness across the paper width owing to non-uniform nip pressure or application amounts;
  - drips and stripes in the starch coating;
  - patches where the coating is not correctly applied;
  - orange peel appearance.

  All of the following appearance factors can be improved by adding a proportion of the dye to the stock as well, providing they are not too serious.
- Bleedfastness: the bleedfastness of size-press dyeings can be poorer than that for stock dyed papers of the same depth of shade, since all of the dye is concentrated at the surface. The solubility of the starch coating can also mean that the dye attached to the starch can be removed along with the starch in the bleedfastness test. This can be improved by the addition of a wet-strength agent to the starch [13].
- Bronzing: described above in Section 9.2.2.
- Colour corrections, colour changes and clean-down: depending on the total volume of the size-press service system and the addition point of

the dye any changes made to the dye dosage volumes, although initially seen rapidly, are not complete for a significant time, owing to the time taken for the size-press system to reach equilibrium. This time tends to be greater than that for normal continuous dyeing. For the same reason, colour changes are also more difficult than with continuous dyeing, even if the production is planned so that clean-downs are not required between colours. For example, starting with yellow then moving through orange to red. Once the coloured production is over, the size-press service system and the size press itself must be cleaned down and the remaining coloured starch solution must be dealt with according to the mill's internal systems. Where split addition is used, involving the addition of some of the dye continuously in the stock, this can be used to control colour to some extent, providing the changes are not too large.

The disadvantages of size-press dyeing make it unsuitable as a replacement for continuous dyeing in most cases. In particular, the problem of furnish and process variation leading to absorbance differences, and therefore colour differences, is a difficult one to overcome. For example, size-press dyeing can produce excellent results and then at a later time, when the amount or type of broke used is different, the results can be disappointing.

9.2.5 Split addition

The successful use of size-press dyeing alone is quite difficult. Aside from those factors already described, the effect of coloured broke is also very important. Broke from the current production grade which is feeding back into the stock will be coloured, and the colour variation produced will be difficult to control at the size press. For this reason, apart from special cases such as the shading of high whites, for example, size-press dyeing is usually used in conjunction with continuous dyeing in the stock. Under this split addition method, the total amount of dye used is divided between the size press and the stock, with the majority of the dye being added at the size press for economy. The dyes used in the stock need not be the same as those used in the size press, provided the base paper is paler but of similar shade to the final colour required. The use of continuous dyeing in the stock reduces the contrast of the surface dyeing with the base sheet and therefore improves the uniformity of the dyeing. Colour can also be controlled to a greater extent by adjusting the continuous stock addition, rather than the size-press addition, to account for the amount of coloured broke being fed back into the stock.

9.2.6 Colour recipe match prediction

As with stock dyeing, a particular coloured sample to be produced on the size press can be matched in the laboratory by using a colour match prediction software package and a reflectance spectrophotometer. Initially, a range of concentrations must be applied on the laboratory size press for each dye, to be used as calibration dyeings in the matching process. The preparation of the calibration dyeings is extremely important to the accuracy of the results produced for matchings. Recipe match prediction results from the laboratory can be used as a guide when matching the shade on the papermachine. This process saves machine time by isolating the specific dyes and the approximate amounts required. This is particularly important in size-press dyeing as colour corrections take longer than with continuous stock dyeing. The laboratory calibration curves can also be used to estimate the saturation levels of each dye in the size press as well as which dyes are susceptible to bronzing.

9.2.7 When should size-press dyeing be considered?

There are cases when size-press dyeing may be the most economic method of dyeing paper, or the only way to produce a particular shade. In general, continuous stock dyeing should be considered as the usual route to producing most coloured papers; however, there are distinct applications where size-press dyeing may come into its own, for example:

- where uniformity is not a problem and the contrast with the base paper is not great because of the relatively dark colour of the furnish – this occurs in packaging grades such as testliners, for example, where the requirements for colour control and uniformity are not necessarily too great and the base colour of the furnish is close to that of the final product;
- in the production of deep shades where there is a clear cost advantage to using size-press dyeing compared with stock dyeing – this is particularly the case where the substance is over 80 g m^{-2}: split addition should usually be used in this case;
- in cases where continuous dyeing is not available – in this case dye can be added in a batchwise manner to the size-press solution and the colour change times are often faster than those in batchwise stock dyeing at the pulpers or machine chests.

It should be noted that bleedfastness may or may not be important. Bleedfastness problems can arise in subsequent processing stages of the paper, for example, gluing, or coating. Alternatively, bleedfastness properties may be irrelevant if the paper is to be sealed beneath a clear laminate layer or coating.

9.3 Dyes used in on-machine coating

Dyes are added to coating slips in a variety of applications, some of which can be carried out on-machine. Dyes are generally added as shading dyes for high whites or in relatively pale shades such as those used for office papers and carbonless copy papers. Where the paper is used in food-contact grades the coated side is generally not in contact with the food itself. If it is, it should show no colour bleeding under the British Standard BS EN 646:1994 test [14], if the paper is required to meet the German Federal Health Office requirements [15]. Lightfastness can be important depending on the end use of the paper or board. Since the exact coating recipes are many and varied they will not be discussed here; however, the types of dyes and pigments used and their applications follow.

9.3.1 Dye and pigment classes

Anionic direct dyes are usually compatible with all coating slips. They have only a low substantivity for the calcium carbonates, clays and latices generally present but they have a strong affinity for the binder materials present such as starch, CMC, PV alcohol and casein. They can be added at any point in the preparation of the coating slip provided they are well mixed before use. They are not compatible with strongly cationic additives but these are not normally used in coating recipes. Lightfastness is relatively good, as is bleedfastness. Bleachability is also good in most cases.

Cationic direct dyes are not compatible with normal coating slurries. A special addition process must therefore be used such as those provided by Clariant [16]. When used correctly they can show many advantages such as good bleedfastness, good bleachability and economic coloration, since they are strongly substantive to the chalks and clays generally used.

Cationic basic dyes are substantive to the chalks and clays typically used in the coating process; compatibility with the coating slip should always be checked before use. Lightfastness is very poor.

Pigments can have higher bleedfastness and lightfastness than the soluble dyes used in the paper industry. This is not always the case, though, as some individual pigment types have very poor lightfastness. Since they have no substantivity for the components of the coating slurry they are simply entrapped physically inside the coating on drying. They are compatible with most coating slip recipes. One problem that is occasionally encountered with pigments used in paper coatings is that they are usually supplied as suspensions which can settle out over a period of time. They must therefore be stirred using a suitable high-speed stirrer before use to redisperse the pigment particles. If not stirred correctly then pigment agglomerates can end up in the coating slip causing spots and streaks of colour in the final coating. Therefore, the use of a suitable screen filter is advisable to remove any larger agglomerates before entry to the coating slip. Fluorescent pigments for coatings are also available which can be used in security and stamp papers. Pigments for paper coatings tend to be expensive compared with dyes and cannot be bleached.

Because of their high solubility and therefore lack of substantivity and very poor fastness properties, acid dyes are not normally recommended for coating use.

The use of dyes and pigments in the coating slurry may generate foam, in which case a suitable anti-foam agent should be used that will affect neither the stability of the dye or pigment used, nor the runnability and performance of the coating slip.

9.4 Calender stack dyeing

In some situations dye is applied to the surface of the paper at the calender stack. This is achieved by adding dye to a water box which is present to moisten the sheet. The dye is therefore transferred to the paper in solution and the dye is absorbed quickly by the fibres. Good results can be achieved only if the dye is sufficiently substantive to the fibres. Pick-up is limited owing to drying capacity before reel-up, so only very pale shades can be produced by this method.

9.5 Spraying

Dyes can also be added to the surface of the paper by spraying a solution of dye onto the newly formed wet sheet before pressing; this system is generally only used when the requirements are not particularly strict regarding levelness and appearance.

9.6 Broke treatment

In all operations where coloured paper is produced, whether it is surface coloured or not, the problem arises of what to do with any broke left over once the colour run has been completed. There are generally two options open to the papermaker – storing for later use, and bleaching – and circumstances dictate which is the most appropriate. Of course, other additives may be present which require separate treatment such as wet-strength agents and coatings.

Storage is the simpler option of the two, whereby the coloured paper is stored until the same colour, or a similar colour, is made again and is then re-used at that time. This saves on dye usage as the colour of the broke contributes towards achieving the shade. It also negates the need for a bleaching operation; however, a suitable storage area is required along

Table 9.2 Bleaching conditions

Bleaching agent	Active bleach (%)	Temperature (°C)	pH	Time[a] (mins)
Hypochlorite	Chlorine: 1–2	50	4.0	30–60
Hydrosulphite	Hydrosulphite: 1–2	60	6.0	30–60

[a]In general, the higher the consistency at which bleaching can be carried out the faster the results.

with an effective broke management regime. In some regions this is the only method used for recycling coloured broke, although this system is not feasible if the shade range is particularly wide.

With respect to bleaching, most of the dyes used in the paper industry are bleachable to a greater or lesser extent, but pigments are usually not. As long as dyes are chosen that are bleachable under the conditions used, coloured broke can be bleached back to white, usually in a separate bleaching pulper or indeed a dedicated bleaching plant.

The bleaching agents most commonly used are those previously mentioned: hypochlorite, hydrosulphite or dithionite and FAS. The efficiency of the bleaching process depends on the conditions available. Generally, the bleaching should be carried out at a temperature above 50°C (though not with hypochlorite), particularly when the depths of shade involved are medium to deep. If the broke is to be overdyed then total colour removal is not always necessary. Once the bleaching is completed the bleaching chemicals remaining must be de-activated by a suitable reagent, such as anti-chlore in the case of hypochlorite, or they will continue to bleach any dye added later in the system, thereby increasing dye usage and costs.

The most favourable conditions for bleaching in each of the two most common cases (hypochlorite and hydrosulphite) are shown in Table 9.2. Note that on no occasion should bleaching agents be used together or mixed in any way. The supplier's health and safety instructions should always be strictly adhered to. In some countries, existing legislation concerning the materials found in the effluent prevents the use of hypochlorite bleaching. The conditions shown in Table 9.2 have been found to be the ideals and are used in the laboratory; however, it is not generally practical for the papermaker to use some or all of these conditions. Bleaching can still be used at pH 7 and at low temperatures (30°C–40°C), provided the particular dyes are easily bleached and the shades are pale.

9.7 Survey of surface dyeing

A survey was done of surface dyeing practice in a number of countries (Argentina, Austria, Belgium, Brazil, Canada, Chile, Colombia, Eastern

Europe, France, Germany, Holland, Japan, Mexico, New Zealand, South Africa, Spain, Switzerland, Thailand, United Kingdom, USA). The resulting estimated information is as follows.

- Distribution of application methods: the total number of paper mills using surface application was 250–300, and of these the breakdown of the methods used is shown in Figure 9.3(a).
- Size-press dyeing and split addition: the total number of mills using size-press dyeing was 110–150, of these the ratio of size-press addition alone compared with split addition is shown in Figure 9.3(b).
- The uses of size-press dyeing: the relative importance of the different uses of size-press dyeing are shown in Figure 9.3(c). The category 'other' shows those which use the size press to achieve coloured shades not included in the other categories, for example, deep colours.

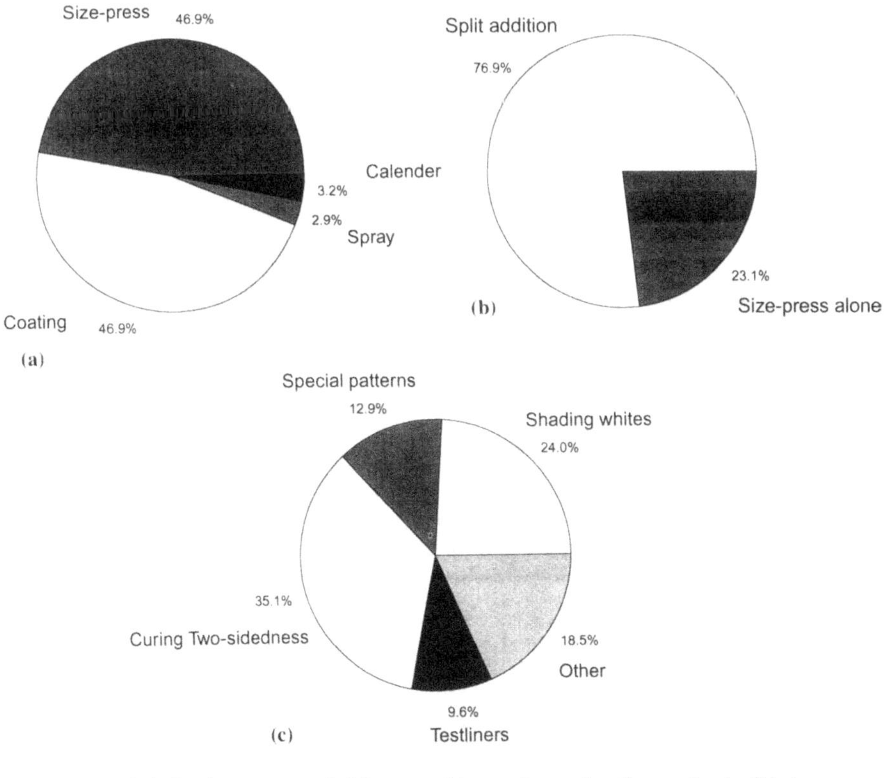

Figure 9.3 Relative importance of: (a) on-machine surface coloration methods; (b) size-press dyeing alone and split addition; and (c) size-press coloration usage.

9.8 Conclusions

Surface coloration is a small but commercially important field of application. A significant number of paper mills use surface colour application successfully to achieve results which they cannot achieve by other methods alone. Size-press dyeing is the most important of these methods, however; spraying and calender stack dyeing are only rarely used. Coating applications also use colour; however, the coating is not used as simply a method to apply colour to the paper. Although there are problems associated with size-press dyeing, the number of paper and board producers using this method today shows that results can be successful, providing the process is well understood. It should be noted that in all of the cases mentioned above pre-trials should be carried out in the laboratory to determine compatibility of any product with the other additives, the process, the equipment and the materials already in use, before proceeding to bulk trials.

Acknowledgements

Trevor Berry, Nick Dunlop-Jones, Ade Forster and Jane Tindal are thanked for their help in preparing this chapter. Clariant affiliated companies worldwide are thanked for providing useful information.

References

1. Sandoz (1995) The dyeing of paper and board in the size press, Sandoz paper technical booklet SB 5944/75; copy available from Clariant UK Ltd, Calverley Lane, Horsforth, Leeds LS18 4RP.
2. Schwab, O. and Rapp, H.-U. (1974) Investigations into surface dyeing and sizing in the size-press. *Papier*, **28** (10A), V43–51.
3. Loewe, H.D. (1974) Using synthetic sizing agents in paper manufacture. *Paper Technology*, **15** (3), 158–63.
4. Martin, G. (1974) Dyeing of paper and board on the size-press. *Wochenbl. Papierfabr.*, **102** (11–12), 429, 432–4.
5. Sugimoto, T., Komai, A. and Enokizono, S. (1977) The size-press surface dyeing of paper with anionic dyes. *Kami Pagikyoshi*, **31**, (6), 49–61.
6. Vidal Jane, J. (1973) Size-press dyeing process. *Asoc. Mex. Tec. Ind. Celul. Pap. [Bol.]*, **13** (5), 349–56.
7. Etlis, N.M. (1973) Modern trends in paper sizing. *Asoc. Mex. Tec. Ind. Celul. Pap. [Bol.]*, **13** (5), 357–68.
8. DeRemigis, J., Martin, J.C. and Hamilton, J. (1974) Closed loop colour control at the size-press. *Instrumentation in the Pulp and Paper Industry*, **15**, 57–64.
9. Micallef, J. (1975) Use of the size-press for dyeing paper (in Italian) *Ind. Carta*, **13** (2), 47–50.
10. Davenport, B. (1974) Abitibi automates size-press colouring with EA poly-sensor. *Pulp Pap. Mag. Can.*, **75** (5), 30–1.
11. De Remigis, J. and Hamilton, J. (1974) New optical sensor opens way to on-line computer control of colour. *Pulp and Paper*, **48** (12), 65–8.

12. Murray, S.G. (1996) Dyes and fluorescent whitening agents for paper. *Paper Chemistry*, 2nd edition (ed. J.C. Roberts), Blackie Academic and Professional, London, pp. 161–93.
13. Arnold, E. and Martin, G. (1994) Possibilities and limitations of size-press dyeing. *Wochenbl. Papierfabr.*, **122** (9–10), 381–91.
14. BSI (1994) BS EN 646:1994 Paper and board intended to come into contact with foodstuffs: determination of colour fastness of dyed paper and board, British Standards Institute, London.
15. BGA 36th Recommendation of the Bundesgesundheitsamt (Federal Health Office) of the Federal Republic of Germany, Berlin.
16. Coating slip dyeing – new special process with Cartasol® K dyes. Sandoz paper technical information bulletin 6031.28.84, 2 July 1984; copy available from Clariant UK Ltd, Calverley Lane, Horsforth, Leeds LS18 4RP.

Further reading

Chekunina, L.I., Dikler, Ya.E. and Strel'nikova, K.K. (1988) Surface bleaching of boxboard. *Bum. Prom-st.* (1), 13–14.

Glittenberg, D. (1993) Size press–film press. *Allg. Pap.-Rundsch.*, **117**, (48), 1264–1269.

Klein, H. (1986) Dye-preparation and metering method for the dyeing of paper and plastic sheets. *Coating*, **19** (7), 236–8.

Klein, H. (1994) Coating technology in the production of printing papers and folding boxboard (Part 1). *Coating*, **27** (6), 206–8.

Lambert, B. (1989) Surface treatment with the size-press – a lot of potential. *Pap. Technol.* **30** (7), 10–11, 14.

Palenik, K. (1994) Paper colouring with water-soluble dyestuffs. *Przegl. Papier.*, **50** (3), 98–9.

10 The surface treatment of fine business papers

M. WATSON

10.1 Introduction

In recent years, the traditional methods and materials used for surface treatment of (fine) business stationery papers have come up against the extra demands of modern technology. Size-press impregnation of such papers with starch is typical of normal surface modification and gives as an end result a paper surprisingly tolerant of older printing techniques, such as lithography and dot matrix printing: so-called 'impact' printing.

However, one of the major developments to affect this segment of the paper industry in recent years, particularly since the mid-1980s, has been the advent of desktop publishing. This has brought revolutionary changes to the way in which documents are produced. Publications can now be made from in-house equipment placed on the office desk or in the home.

Digital information from a personal computer can be quickly and easily turned into hard copy using the now ubiquitous laser or ink-jet printer; so-called 'non-impact' printers. The print head does not contact the paper – hence the term non-impact. The growth rates of installed printers and associated cut sheet consumption have been very high and this situation is likely to continue for some time.

These major changes in the business and office papers market have exposed deficiencies in the traditional surface treatment materials and application methods. Generally speaking, paper surface attributes required for optimum laser print performance are quite different from those required for ink-jet printing, and simple starch treatment is not necessarily best for either. The paper industry has responded to these demands by the use of some quite sophisticated surface sizing chemicals and high-performance pigments. These materials may be applied at the size press or more effectively by coating or film-press technology in order to locate them more readily at the paper surface.

10.2 Non-impact printing

10.2.1 Laser printing

Laser printers use essentially the same technology as photocopiers, but the former produce originals from digital information whereas the latter produce copies from an original (Figure 10.1). A photosensitive drum is charged across its surface with a pattern corresponding to the image to be printed. The charged areas pick up the imaging material, called toner, which is transferred to the paper by physical contact. Toners are typically composed of finely divided particles of thermoplastic polymeric resins, for example, styrene–acrylic copolymers, as well as minor amounts of waxes, charge control agents, and carbon black as the pigment. After transfer to the paper, the toner particles are fixed to its surface by fusing under pressure with heated rollers. Some printers fuse the toner by pressure alone, but this is rare. Temperatures in most laser printer fusers are around 200°C.

10.2.2 Ink-jet printing

To date, there are three distinct classes of ink-jet printer.

- Continuous drop: these printers are mostly used in packaging applications and for pre-press proofing; they typically use solvent-based inks.
- Solid ink-jet: these are rare at the moment, being used for personal printers, particularly for colour work; the inks are fusible solids and rather similar to the toners used in laser printers.
- Impulse drop: for business stationery this class of printer is by far the most common, and for this reason the discussion that follows refers to these printers only; there are two types of printer in this class:
  - piezo-electric (PE);
  - thermal.

Both impulse drop systems operate an ink-drop-on-demand system whereby digital character input from computer to printer stimulates a drop generator to fire ink drops as and when required. In PE devices, the drop generator is a crystal that expands and contracts under the influence of an

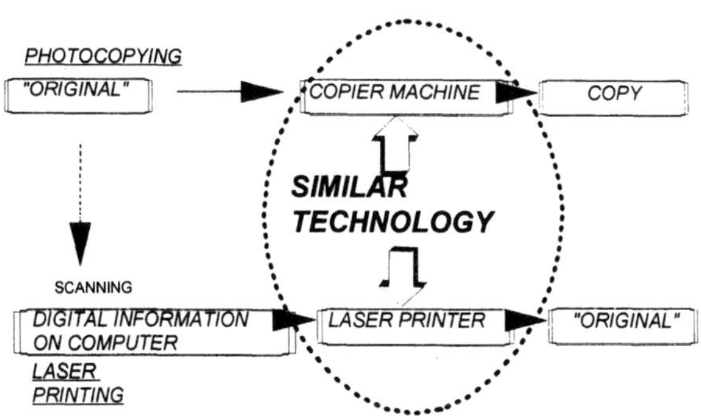

Figure 10.1 Comparison of xerographic copiers with laser printers.

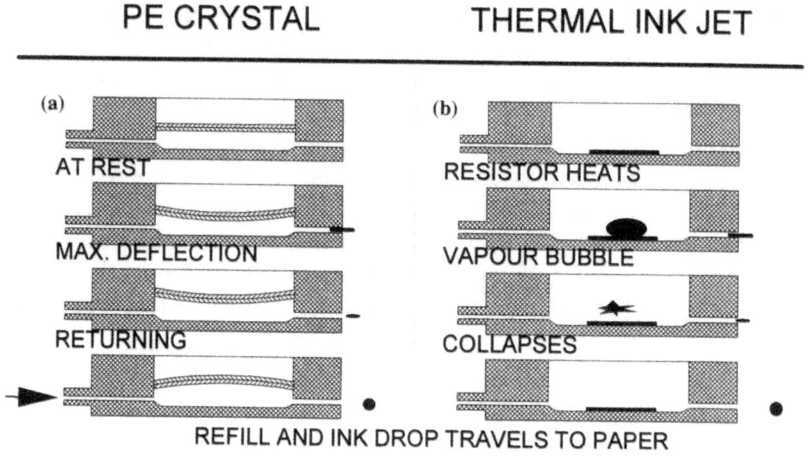

Figure 10.2 Impulse drop ink-jet printers: (a) Piezo-electric crystal ink-jet printer; (b) thermal ink-jet printer.

electrical stimulus supplied by the crystal driver. The PE crystal reacts to the input and fires ink droplets out of the nozzle when required. In thermal machines, the input signal generates an electric current through a resistor inside the nozzle. The heat produced locally boils the ink, thus producing a vapour bubble which forces an ink drop through the nozzle at high speed as a result of pressure increase. The impulse drop principle is shown in Figure 10.2.

Compared with continuous drop, impulse drop is simple, inexpensive, quiet, and economical on ink; conversely it is relatively slow, limited with respect to ink formulation, and large-array printing is not yet feasible.

Ink-jet inks for impulse drop are essentially water-based, also containing minor amounts of co-solvent (such as glycols), dyes or pigments, surfactants and penetrants.

10.3 The chemical natures of non-impact printing image-forming materials

10.3.1 Laser printer toner

Polymers. Styrene–acrylic copolymers used to be the main binder in toner formulations. To take account of new demands on toners, new polymers are being introduced, for example, those based on styrene–butadiene, ethylene, ethylene–vinylacetate, polyester, and epoxy. These advanced toner formulations have produced improved toner hardness, rheological properties to give better fusing behaviour, and allow changes in the overall laser printing process such as cold pressure fusing, radiant and flash fusing.

The polymer glass transition temperature and melt index are critical parameters in determining toner viscosity at a given temperature and indicate at which point the toner softens (60°C–70°C) and becomes fixable.

Charge control agents. The main influences of these materials on the toner properties are charging rate, fuser life and triboelectric (friction-induced) charge. Essentially, they adjust the polarity and electrostatic charge of toner. For positively charged toner formulations these agents are mainly specific black dyes; for negatively charged toners, a more recent development, complex metallic dyes are generally used.

Carbon black. The selection criteria for carbon black include level of blackness, particle size, surface functionality, oxygen content, surface energy and acidity – an acid black is preferred for negative toners and an alkaline black for positive toners.

Other materials are often added in minor amounts to improve flow, photoconductor cleaning and toner transfer.

The particle size of toners is important, and is normally in the range 5–10 microns.

10.3.2 Ink-jet inks

Ink-jet inks for impulse drop printers are currently water-based. The inks contain other substances which are vitally important for proper performance. Co-solvents such as diethylene glycol and *N*-methylpyrollidone are added to retard water evaporation, prevent nozzle blockage and also to control surface tension and viscosity which will have a big influence on print quality on paper.

The presence of water raises the risk of printhead corrosion and bacterial contamination. The inks contain dyes which provide the appropriate colour; it is very important that the dye and resulting ink are pure and free from particulate material. A recent development in ink-jet ink technology is that of pigment-based ink, where the ink manufacturers appear to have overcome the aforementioned problems of particulate 'contamination', and the Hewlett-Packard 660C printer uses a pigmented carbon-black-based ink.

Pigment-based inks offer several advantages over their dye-based counterparts – most notably higher optical density and improved light-fastness. However, the presence of poorly bound pigment means poorer rub resistance than a dye-containing ink. Pigment-based coloured inks are currently under development, and there is some debate as to the extent to which pigment inks will replace dye-based inks.

Applications where high optical density, waterfastness and lightfastness are critical (for example, signage and display) may be where the new technology will be adopted.

10.4 Paper requirements for non-impact printing

Unfortunately for the papermaker, paper requirements for laser printing are very different from those for ink-jet printing. The market appears to expect versatility from the paper, with at least acceptable print quality on a range of print technologies. This contradiction has resulted in vigorous development of novel surface application formulations and hardware, the effects of which will be discussed later in this chapter.

10.4.1 Laser printers

In terms of print character definition and quality, the laser printing process is highly tolerant of paper properties. However, the physical and chemical nature of the paper surface are critical in determining the image permanence. The electrical resistivity of the paper is important in that it affects the way in which the toner is held on the paper prior to fusing; a value of 10^{10}–10^{12} Ω m^{-2} is generally believed to be optimum. Paper resistivity tends to be in this range anyway, and printer manufacturers have designed their hardware to be most effective in this region. Image quality problems caused by (insufficient) conductivity are manifested as print areas with macrovariations in optical density, often with missing or part-missing characters. The toner which has found its way onto the sheet will be well fused to the surface. The level of conductivity can be manipulated by adjusting the moisture content of the paper (higher moisture gives higher conductivity) and by discreet addition of common salt with the usual size-press mix, normally starch.

Perhaps the most common and serious problem associated with laser printers is poor toner adhesion. This is most noticeable when the sheet is folded along an axis containing characters. On most papers, these characters will flake off to leave missing or part-missing characters – a totally unsatisfactory situation.

Although the printer itself or toner can be responsible for poor toner adhesion, the paper is often the culprit. Higher grammage papers (for example, higher than 110 g m^{-2}) may suffer poor toner adhesion purely as a result of the higher substance which dissipates some of the fusing heat.

The two paper properties having most influence are smoothness and surface energy. These factors are the same as those governing the general adhesion rules for liquid–solid interactions [1, 2].

Smoothness. All other things being equal, smoother papers will usually have better toner adhesion than rougher ones. Some care is needed here; a degree of micro-roughness (1 micron level) may be beneficial, but larger variations in surface topography may significantly impair adhesion. Micro-roughness helps the small toner particles key more effectively to the sheet, so enhancing adhesion.

The mechanism for reduction in adhesion due to macro-roughness is fairly simple. The printer fuser roll is unable to make complete and intimate contact with the toner, so fusing is less effective. For this reason, textured papers often have poor toner adhesion.

Surface energy. According to adhesion theory, the surface energy of the adherend (paper) should be higher than that of the bonding liquid for spontaneous adhesion to occur. The toner is softened in the fusing step and flows onto the paper, behaving rather like a liquid, albeit for a very short period of time.

In order to confer hydrophobicity to paper it is internally sized with polymeric or resinous materials, most notably rosin (abietic acid derivatives) and alkyl ketene dimer (AKD). The aim here is to reduce the surface energy of the paper to avoid wetting (by water). It is this act of sizing with low-energy polymers that gives problems with laser toner bonding, as the surface energy of the paper is dramatically lowered. Unsized paper gives excellent toner adhesion but is unsuitable for applications involving water-based inks, notably for ink-jet printing and for writing with fountain pens.

AKD sizing is particularly unfriendly to toner; this, coupled with a propensity amongst papermakers to add too much, can lead to seriously poor toner adhesion. Table 10.1 shows comparative results for typical AKD and rosin sizing systems.

Traditionally, starch is applied at the size press as a surface size to mask the detrimental effects of internal sizing agents. The presence of starch

Table 10.1 Dynamic contact angle (determined by the geometric method), Cobb and resultant laser printer toner adhesion (determined by the tape-pull method) for various papers. AKD = alkyl ketene dimer

Size	Surface tension (dynes cm^{-1})	Cobb (g m^{-2} min^{-1})	Toner adhesion (percentage retained optical density)
AKD with starch	33	24	50
Rosin with starch	39	24	65
AKD with styrene–acrylic copolymers and starch surface size	35	21	55
Reduced AKD with styrene–acrylic copolymers and starch surface size	39	23	70

increases the surface energy of the paper and hence the toner adhesion, though this can be to the detriment of ink-jet printing, which will be discussed in the next section. Chemical compatibility between the paper surface sizing chemicals or coatings and the toner is undoubtedly an important factor in toner adhesion – manufacturers of surface sizing chemicals have noted this and recent developments have attempted to take advantage; this will be discussed in Section 10.5.1.

10.4.2 Ink-jet printers

Desktop laser printers are mostly monochrome only, although the first colour machines are now hitting the market. For ink-jet, monochrome and colour devices have been around for a good few years. Today, colour ink-jet machines can be purchased for a few hundred pounds sterling. Consequently, the growth in colour ink-jet placements and sheet usage has been very strong.

Paper requirements for good monochrome print quality are actually quite straightforward; the arrival of colour capability with more complex inks has made life difficult for the papermaker. Unlike in the laser printer, the aqueous inks used in impulse drop ink-jets are highly paper dependent; the inks are prone to absorbing into the paper [3]. Uncontrolled absorption may lead to problems, some of which are schematically shown in Figure 10.3. For good print quality, minimal penetration and lateral spread of the ink are desired. Problems may occur if:

- the ink spreads excessively, giving feathering or wicking;
- the ink penetrates too deeply, and dye or pigment is lost into the paper, giving dull print;

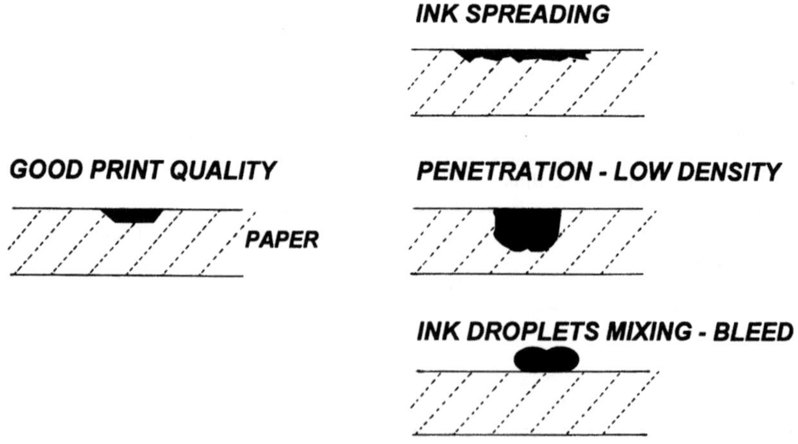

Figure 10.3 Print quality comparison.

- one or more inks penetrate too slowly, giving either ink bleed or slow drying.

Most inks contain dyes and this, together with the occurrence of penetration into the substrate, means that adhesion of the ink is seldom a problem.

Monochrome ink jet. The most important paper property for monochrome printing is hydrophobicity brought about by internal sizing. The aim here is to reduce the surface energy of the paper to well below the surface tension of the ink. Table 10.2 shows typical values for the surface tension of some common inks. The addition of surfactants and penetrants to the ink suppresses the surface tension compared with water [3].

An unsized sheet with high surface energy may have excellent laser toner adhesion but would give a very poor result with today's ink-jet printers because of excessive and rapid ink absorption. The ink would dry very quickly, but the price to pay would be dull, feathered images probably with ink strike-through to the non-imaged side.

A hard-sized sheet would give opposite print attributes: crisp, dense characters but with slower ink dry-time. The mechanism for ink drying will be a mixture of absorption and evaporation, depending on the paper hydrophobicity, the ambient temperature and humidity.

Colour ink jet. The advent of colour has complicated the ink-jet ink–paper interaction scene considerably. For monochrome ink jets, we need only to consider one interaction, that of black ink with paper. For colour, we have several different ink–paper interactions pertaining to different colours as well as ink–ink interactions occurring at the paper surface before and after the inks have been fully absorbed into the paper.

The basic ink colours are magenta, cyan and yellow: to obtain a wider range of colour, composite colours are used with one colour overlaying another on the paper. Different coloured inks are formulated differently and have varying surface tensions. This makes controlled absorption very

Table 10.2 Surface tension (determined by the platinum ring method) of some aqueous dye-based ink-jet inks

Ink	Surface tension (dynes cm^{-1})
Black	39
Cyan	48
Magenta	48
Yellow	47

difficult to achieve with conventional papers having internal sizing and starch surface sizing.

The most common print problems associated with colour ink jet are dull, uninspiring colours caused by excessive ink absorption, or bleeding of adjacent inks into one another, manifested as ragged borders between areas of adjacent colour. The former will be accompanied by fast ink drying, the latter by much slower ink drying but more vivid colour.

Bleed occurs when two (or more) colours run into one another before they are absorbed into the substrate. Ink bleed clearly cannot occur with monochrome printing.

Another common problem encountered is mottle of block colours. This is a wetting phenomenon where the ink compatibility with the paper is low in the thermodynamic sense. In this case, the contact angle will be high, leading to reticulation of the ink on the paper surface. It is this beading effect which leads to mottle. This is a common phenomenon in hard-sized papers.

AKD-sized papers tend to give bright colours but slow ink drying and hence ink bleed. In this case a surface application will be necessary to modify the surface wetting properties: conventional starch applications are effective in that they shield the ink from the powerful hydrophobicity of the AKD. Reduced ink bleed and faster drying will result, but at the expense of colour brightness.

Rosin-sized papers generally give a compromise in print quality: little ink bleed with moderately bright colours and fairly fast dry-times. Consequently, starch surface sizing does not always appear to be as effective in improving print performance with these papers.

The difficulty in achieving good quality colour ink-jet prints with plain, uncoated papers has been recognized by the market and the growth of colour ink jet has been closely shadowed by growth in papers dedicated to colour ink jet. Such papers are coated with ink-receptive chemicals and pigments to absorb and immobilize the ink within the coating. These products will be covered in more detail in Section 10.5.2.

10.5 Surface treatment for non-impact printing papers

It is clear, then, that for laser printing applications a soft-sized sheet is preferred with respect to toner adhesion, whereas for ink-jet printing a harder-sized sheet gives better print quality. It is a very difficult task consistently to get the sizing compromise just right. One fairly common approach is to obtain the hydrophobicity using a material with more compatibility for toner than the internal sizing chemicals, or with less propensity to depress excessively the surface energy. Such materials are usually known as polymeric surface sizes. Another approach is to apply a very thin, say 5–20 micron, layer of water-sensitive material onto a hard-sized or hydrophobic substrate. This material can be polymeric, or a hydrophilic pigment (for example silica) or indeed a combination of both. This latter approach is most effective when the material is located at the paper surface by either coating or film-press application; size-press impregnation is much less effective as the active materials are lost into the base paper or distributed unevenly over the surface.

10.5.1 Polymeric surface sizing agents

An increasing number of paper chemical suppliers now offer a range of these materials, with many dubiously claiming to be the panacea for all the ills of non-impact printing. To be sure, there is no doubt that some of these materials can turn a borderline situation into an acceptable one. These chemicals can be broadly classified as follows:

- styrene–maleic anhydride (SMA) copolymers;
- styrene–acrylate (SA) copolymers and terpolymers;
- polyurethanes (PUs).

The SMA copolymers are quite old technology which have been largely superseded by the more advanced SA and PU polymers. The SMAs are normally supplied as quite viscous solutions, or more accurately as colloidal dispersions. The SAs are usually in the form of micron-size emulsion droplets in water; these have low viscosity and are easy to pump and generally handle. PUs often come as colloidal dispersions of low viscosity.

The degree of hydrophobicity varies according to the chemical type and structure; for example, SAs of differing structure can have different hydrophobicity. This will have a considerable effect on the paper properties for non-impact printing and on the addition rate of the polymer. In the author's experience, the SMAs are least hydrophobic and the PUs most hydrophobic. These materials are applied at the size-press, blended with the regular size-press starch [4].

Addition rates do vary, but a typical figure is 5 wt% polymer for 95 wt% starch on a dry basis. The object is to locate these polymers at the paper surface for maximum efficiency. With conventional size pressing this is difficult to achieve, although improvements in ink-jet quality and toner adhesion usually result. Use of these polymers gives a boost to sizing, and a small reduction in internal sizing chemicals is normally possible. Table 10.1 shows the surface energy (measured by dynamic contact angle) for AKD-sized paper having a normal starch surface size, and one having a small amount of styrene–acrylic copolymer with the starch and a reduction in the addition rate of AKD. Laser toner adhesion values are also shown together with Cobb(60s) sizing figures. SA copolymers have some chemical compatibility for the resins used in the toner itself, which are thermoplastic and often styrene-based. This helps adhesion. The surface energy of the polymer-treated paper is higher, although the Cobb value is similar. The treated paper ink-jet quality for monochrome was unchanged, but colour prints showed slightly reduced ink bleed and print mottle and marginally faster dry-time.

10.5.2 Coated ink-jet papers

Uncoated business papers will deliver reasonable colour ink-jet quality for simple applications, but for complex graphics where highly vivid and sharp mottle-free images are necessary a dedicated paper grade is required. Such grades are coated on at least one side with an ink-receptive layer typically at a coat weight of about 5–15 g m^{-2}. This layer has the task of rapidly absorbing the ink solvents (water plus any cosolvents) and ensuring the ink dyes or pigments are prevented from penetrating into the base paper. This rapid drying and location of dyes at the surface gives bright colours with no mottle or ink bleed. This is an analogous situation to conventional pigment-coated papers used for offset litho printing [5].

The fine surface pigments absorb the mineral oil fraction from the ink, leaving the high viscosity binder oil and pigment on the surface of the coating. The result is a faster drying and more intense image than if an uncoated paper were used. (Although the mechanisms are parallel, for ink-jet printing the pigment and binder should be hydrophilic and for litho printing the coating should be hydrophobic.) For offset litho, the ink-receiving layer is hydrophobic, whereas for ink-jet printing the ink-receiving layer is hydrophilic, with the most common ingredients being silica as the pigment, and minor amounts of partially hydrolysed polyvinylalcohol (PVOH) as the pigment binder. Other pigments such as chalk may be present, as well as additional binders, commonly polyvinylpyrollidone (PVP). The net result is almost instantaneous absorption of the ink vehicle with the coloured dye or pigment remaining at the surface.

The base paper is important. Until recently, these tended to be soft-sized to promote ink absorption. However, inks sometimes penetrated too quickly and deeply, resulting in strike-through to the non-imaged side. Some coated ink-jet papers today are produced on hard-sized base. The ink is restricted to the ink-receiving layer and cannot penetrate much into the base.

The difference in the penetration dynamics of uncoated and coated papers can be studied with various pieces of hardware, such as the Bristow Wheel or Hercules Sizing Tester [4, 6–8] and now with a relatively new piece of equipment, the dynamic penetration meter (DPM). This uses ultrasound to follow the progress of liquid penetration into the paper;

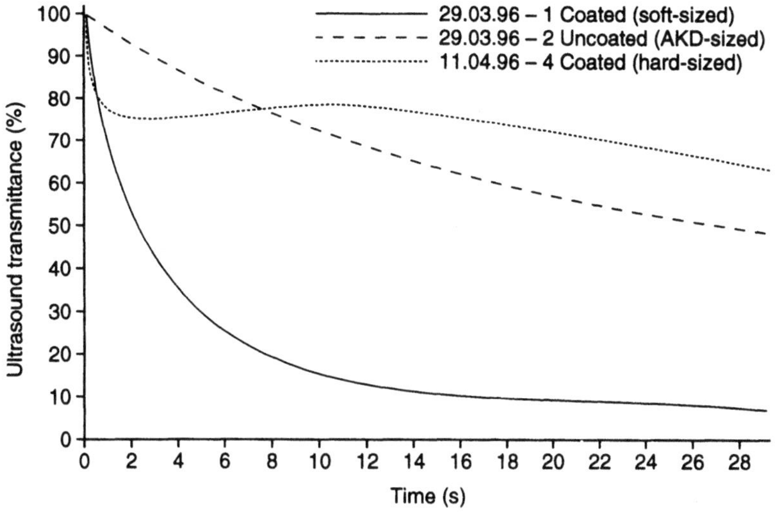

Figure 10.4 Dynamic penetration meter (DPM) graph of ink-jet papers and plain papers.

ultrasound transmission reduces as the paper is penetrated. Figure 10.4 shows a DPM graph for specialist coated ink-jet paper (with soft-sized and hard-sized coating base) and a typical uncoated AKD-sized/starch surface-sized business paper.

The uncoated grade has a definite penetration delay (actually 444 ms) manifested as an initial increase in ultrasound transmittance. This delay is a result of exclusion of air at the paper surface as the water overcomes the starch surface sizing. Once over this, the strong water-repellent effect of the AKD retards penetration, as indicated by a slow reduction of ultrasonic transmission with time.

In contrast, the coated papers have no such penetration delay because of the hydrophilic nature of the coating, and ink absorption will be rapid. With the soft-sized base, penetration continued unchecked through the base and the paper is effectively saturated after 20 s. This paper exhibited ink strike-through with colour printing.

The hard-sized base has no penetration delay either, but after the initial absorption phase penetration is slowed as the interface between coating and water-repellent base is reached. The result is high print quality with no strike-through.

10.6 Surface application methods

For optimum results with either non-impact or impact printing methods it is desirable to keep the dyes, pigments or toner close to, and preferably at, the paper surface. This is especially important for liquid inks such as for offset litho and ink-jet printing. It is also preferable to locate polymeric surface sizes and high-performance pigments where they will achieve most benefit: that is, at the paper surface. Because of this, the method of application of the surface treatment is highly influential on print quality. The traditional treatment method of size-press impregnation is imperfect, yet coating usually involves an off-line process with added costs (assuming you have a coater at your disposal!). These deficiencies coupled with the drive for added value and top performance on the ever widening range of print technologies have spurred the development and growth of film-press technology.

The size press has been around for a much longer time than the relatively untested film press and clearly has a useful role to play: the drawback with a long track record is that the problems become more apparent and well documented. With this in mind, it is interesting to contrast the two techniques.

10.6.1 Size press

The size press basic configuration should be familiar to readers; it is not within the scope of this chapter to cover this aspect in much detail. The equipment itself can take on three general configurations; vertical, inclined or horizontal. Whichever is used, the principle is the same. The press itself comprises two rolls pressed together to form a nip, which is flooded with the surface size mix. One roll is hard, typically metal or hard rubber, and the other is softer, again made from rubber. The size press was originally developed to apply starch size to paper in order to improve surface strength and resistance to penetration by water-based materials, but its use has now been extended (not very successfully) to pigmenting for enhanced functional properties or as pre-coat for subsequent topcoating.

The major factors influencing pick-up are:

- sheet moisture content before the size-press – absorption of the mix is higher at higher moisture contents;
- mix solids content and viscosity – these are the easiest to manage, although the scope for high solids is limited by a need to run viscosities preferably below 50 MPa s; to minimize after-drier steam demand, solids should be maximized;
- level and type of internal sizing – AKD-sized paper will probably be incompletely cured entering the size press and will tend to pick up more surface size; rosin or ASA (alkenyl succinic anhydride) sizing will be substantially cured and hence less absorbent; these factors will also affect the distribution of size throughout the sheet;
- sheet roughness – a rougher sheet will pick up more mix in the surface depressions and valleys than will a smooth sheet;

- machine speed – lower speeds increase pick up;
- nip pressure and roll hardness – higher pressure and harder rolls will reduce pick-up.

In practice, the pre-size press sheet moisture content is kept at around 2% (or lower) to ensure evenness of pick-up through and across the web. This leads to higher costs in extra drier steam demand.

Notwithstanding the above it is difficult to control these variables precisely enough to achieve a homogenous product in terms of surface size distribution; this will induce variability in its end-use functionality. In addition, one is limited in the amount of treatment that can be applied by viscosity; although there are bound to be specific exceptions it is unusual to run at a solids content higher than 15%, indeed 5%–10% is more normal. The low solids means that the usual coat weights possible are only in the range 0.5–2.5 g m^{-2} per side.

Because of the high level of web saturation with a size press it is difficult to run weak or low basis weight sheets without costly and irritating web breaks; it is possible but there are limits on speed.

Pigmented mixes are prone to dusting. This is a phenomenon caused by poor binding of the pigment to the base paper. With the low-viscosity mixes used at the size press, pigment binders such as starch, PVOH, or synthetic latices are all too easily preferentially absorbed into the base, leaving the pigment weakly bound to the paper surface and easily dislodged.

10.6.2 Film press

Several manufacturers offer various versions of this hardware; the principle of operation is similar in all cases and is covered in detail in Chapter 2 of this book. The objective of the film press is to apply more of a 'surface' treatment rather than an 'impregnation/surface' treatment. The main feature which allows it to do this is the ability to run much higher solids and viscosity than can the size press. This enables fairly thick films to be applied which become immobile relatively quickly for a given drier input and are more likely to remain at the surface. Pigment dusting will be lessened as the higher mix viscosity will retain binders in the drying coating until immobility is reached.

The typical solids content range is claimed (by the manufacturers) to be 10%–60%, with attainable coat weights of 2–10 g m^{-2} per side. The higher solids will enable a reduction in drier energy demand. The metered size press should be able to run weak or low grammage webs owing to less penetration of mix vehicle, that is, water; and allow higher machine speeds.

For non-impact printing papers in particular, the ability to apply a light-weight coating of speciality polymeric sizes and pigments on-line and more definitely at the surface may be a cost-effective way to overcome decisively the conflicting paper requirements of ink-jet and laser printing by reducing dependence on the base.

10.7 Summary

The high growth rates in paper consumption for new printing technologies, particularly laser and impulse drop ink-jet printers, have placed extra demands on fine papers for business stationery.

Unfortunately for the papermaker, laser printing prefers soft-sized paper for enhanced toner adhesion whereas ink-jet printing is more paper dependent and generally speaking harder-sizing is preferred to minimize ink absorption. To a limited extent, application of polymeric surface sizing agents or pigments blended with conventional starch at the size press has alleviated this conflict, but speciality pigment coated papers are required for high-quality colour ink-jet printing. With the size press the polymers or pigments cannot be distributed evenly across or through the paper and indeed much is lost into the paper. A possible option for 'enhanced' products carrying light-weight coatings of these speciality chemicals lies in the use of film-press technology. This could reduce dependency on the base paper and give simultaneous optimum performance in laser and ink-jet printing.

Acknowledgement

The author wishes to thank the directors of Arjo Wiggins Appleton for permission to publish the material used in this chapter.

References

1. Patton, T.C. (1970) A simplified review of adhesion theory based on surface energetics. *TAPPI Journal*, **53**(3), 421–30.
2. Lepoutre, P. (1986) Adhesion in paper products. *Svensk Papperstidning*, no. 1, 20–1.
3. Carreira, L., Agbezuge, L. and Gooray, A. (1992) Rates of aqueous ink penetration into papers and their effects on printability. IS & T's 8th International Congress on advances in NIP Technologies.
4. Wasser, R.B., Vazquez, I., Proverb, R.J. *et al.* (1994) Contribution of surface and internal size treatments to performance in black and colour ink-jet printing of plain paper. International Printing and Graphic Arts Conference 1994.
5. Bruno, M.H. and Walker, W.C. (1981) *Printing; Pulp and Paper, Chemistry and Chemical Technology*, vol. 4, 3rd edition, (ed. J. Casey), John Wiley, New York.
6. Bares, S.J. and Rennels, K.D. (1990) Paper compatibility with next generation ink-jet printers. *TAPPI Journal*, **73**(1), 123–7.
7. Barker, L.J., Brevard, W., de Pierne, O.S. *et al.* (1994) Surface absorption characteristics

of ASA paper, wetting dynamics using ink-jet ink and aqueous humectant compositions. *TAPPI Proceedings*, 1994 Papermakers Conference.
8. Borch, J. (1993) Surface characterization of communication papers, review paper. 10th Fundamental Research Symposium, 1993; PIRA International, United Kingdom.

Further reading

Houwink, R. and Salomon, G. (1967) *Adhesion and Adhesives*, vol. 2, 2nd edition, Elsevier, Amsterdam.

11 Barrier dispersion coating of paper and board

T. KIMPIMÄKI and A.V. SAVOLAINEN

11.1 Dispersion coating in general

Barrier polymer dispersion coating is understood to mean the coating technique where latex (that is, an aqueous dispersion of fine polymer particles) is applied to the surface of paper or board as such to form a solid, non-porous film after drying. The target of dispersion coating is to achieve a barrier layer against water, water vapour, grease, oil, gas, etc. by environmentally 'friendly' coating.

Dispersion coatings can be fully recycled, which in fact has been one of the driving forces behind the development of the barrier dispersion coatings. Recycling includes repulping, composting and incineration. Thus dispersion coated barrier products, for example, packaging materials, are more easily re-used than typical extrusion coated applications. In addition, loads to landfills decrease as a result of the composting behaviour of dispersion coated barrier materials. Typical barrier dispersions can be burned, but that is not recommended if they include chlorine in any form.

Polymer dispersions consist mainly of water and polymer. Typically used polymers are modified ordinary styrene–butadienes, different acrylates and methacrylates, polyolefins, vinylene acetates, copolymers of these or natural biopolymers. That is why the term barrier polymer dispersion may cause confusion and should be avoided. During the manufacturing process several constituents are added in order to elevate barrier properties or processability. In addition, polymer dispersions may include various amounts of fillers to increase properties such as runnability and cost-effectiveness. Waxes are widely used to hydrophobize the surface.

Polymer dispersions are typically delivered in ready-for-use form. The solids content varies from 25%–70% and the rest is practically water. The viscosity is normally adjusted to 500–1000 mPa s but the dispersions can be diluted with water. The density of utilized polymers is approximately 1 kg dm^{-3}, but the total product density can vary owing to heavy mineral fillers. This also means that it is sometimes misleading to talk about coating weight because it is not necessarily equivalent to the film thickness.

Barrier dispersions must be mixed well before use to make sure that the original distribution of ingredients is achieved. However, the agitation should be done so that air-bubble generation is minimized; as it should also be during the coating operation. This means that free falls, unnecessary circulation and badly constructed piping should be eliminated. The storage time sustained may vary from weeks to several months, depending on the formulation. Polymer dispersions should not be allowed to freeze because of the loss of colloidal stability. Although frozen dispersion is useless, dispersion coated products can be used very well in frozen and chilled food applications.

A barrier dispersion coated product is a combination of many different variables (Figure 11.1). Perhaps the most important factor is the substrate, paper or board, itself. Only a high-quality substrate facilitates a satisfactory end product. The third variable in addition to dispersion and substrate is the coating method and the coating circumstances. All of these factors have an effect on the final barrier, release and sealing properties as well as on the appearance and performance of a possible second coating or printing.

11.1.1 Dispersion coating methods

Polymer dispersions can be applied by using different types of coaters. The coating method can either pick up excess dispersion on the substrate and after that extract it, or a coater can transfer a predetermined dispersion film to the substrate. However, it is important that the method used must produce an even and uniform dispersion film coating on the substrate, without pressing the coating into the substrate. Dispersion coating can be achieved by means of either on-line or off-line coaters. In addition to increased production capacity, the benefits of using on-line coating as

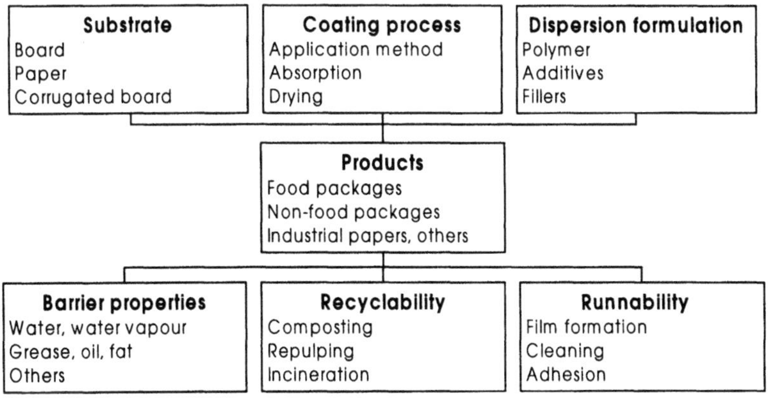

Figure 11.1 Factors affecting polymer dispersion coated products.

opposed to off-line coating are minimized investment costs; and savings in waste, time and labour involved with rewinding.

On-line application is not yet popular although the present coaters are in many cases suitable for dispersion coating; for example, traditional blade coaters are used with good results in off-line application facilities, thus indicating suitability. Another commonly used method is rod coating, where the blade is replaced with a rotating rod, which can be either smooth, grooved or wire-wound. Rod coaters have excellent properties, providing an even film thickness as opposed to an even surface. In principal, an air doctor is one of the best metering methods available, being able to follow almost exactly the contours of the substrate. Unfortunately, it does have speed and coating viscosity limitations.

The film transfer method (metered size press) can be run at speeds in excess of 1000 m min^{-1}, thus being a potentially good coating technique. It may have some problems with detached fibres causing streaks and of dispersion build-up on the backing roll. The metering can be performed by utilizing a blade or grooved rod. The pressure impulse during coating application differs with blade and film transfer coaters, causing different penetration levels of dispersion into the substrate. There have been experiments with size presses for the application of barrier dispersions, but the results were not very encouraging because of film-splitting problems and tendencies toward splashing. However, size presses can be utilized for precoating.

To date, the best results have been achieved by using a separate coating or converting line. In off-line application different types of press have been used with both smooth and anilox applicator rolls. Often, application is performed by the final printing unit of the gravure press when the applicator roll can run either in the forward or in the reverse direction. Other commonly used coating methods are kiss-coaters, puddle-type coaters, nozzle and jet applicators and short-dwell coaters; these are mainly non-modified older types of installation.

Corrugated boards are one of the main products where dispersion coatings are utilized as barrier layers. If the coating is performed as the final operation, the coating methods have to be very gentle because of the layer structure of the substrate. Thus a steel blade may be too hard and should be replaced by a rubber blade. In rubber blade applications the dispersion forms a dam in front of the blade. If one blade cannot wipe effectively enough another blade can be added behind the first one.

After metering there can be a smoothing bar. Some operators find them very useful although others cannot see the benefits as clearly. The idea of smoothing bars is just to touch the wet coating but not meter it. The touching operation may close possible pinholes and orientate the fibres and other material which come up through the coating. The object of using a smoothing bar is to minimize any defects in the dispersion film.

11.1.2 Typically used barrier dispersions

Barrier dispersion coatings can be used both in food and in non-food applications. Food applications require that the components used are mentioned in certain lists of approved materials for food packaging. Commonly used lists are the FDA (Food and Drug Administration of the USA) list, the BgVV (Bundesinstitut für gesundheitlichen Verbraucherschutz und Veterinärmedizin), and the European Communities (EC) directive for materials that can be used in direct contact with dry, wet or greasy food. Moreover, dispersion products must pass migration tests to ensure that harmful contents do not migrate from the coating to the packaged contents before they can be accepted as commercial barrier dispersion products.

The target of the dispersion coating is to achieve a uniform and solid film on the substrate. Figure 11.2 shows the dispersion coating observed from two different viewpoints. In Figure 11.2(a) the surface is defective because there are traces of structure derived from the coating method. Figure 11.2(b) illustrates a good dispersion film in that the holes that can be seen in Figure 11.2(a) are not necessarily deep but only in the top surface. Normally the dispersion coating fills perfectly the valleys between fibres, ensuring excellent adhesion.

Different types of acrylates (the structure of polyacrylate and of polymethacrylate are given in Figure 11.3), styrene–butadiene and copolymers of these are widely used in barrier coatings. Other widely utilized polymers are based on acrylonitrile, fumaric and maleic diesters, olefins, vinyl acetate, cellulose esters and starch ethers. Moreover, various amounts of additives are used so that the total number of ingredients is typically in the range 10–15. Some of these affect polymer characteristics and others influence the colloidal characteristics and the polymerization process itself. Polymer characteristics determined during the polymerization affect, for example, elasticity, glass transition temperature, gel content, etc. Colloidal characteristics affect the flow and stability properties such as particle size, viscosity, surface tension, stability, etc.

Molecular weight is an important characteristic of polymer dispersions. It can be adjusted by chain transfer agents, sometimes called modifiers or molecular weight regulators. However, many dispersions are blends of low, medium and high molecular weight polymers. Thus the polydispersity (the weight average molecular weight divided by the number average molecular weight, M_w/M_n) is in many cases high or even very high. Molecular weight distribution is typically measured by gel permeation chromatography.

Carboxylic acid can be used as an ingredient which increases the ionizing degree of the polymer. Carboxyl groups are formed by the addition of carboxylic acid in the polymerization reaction. The degree of carboxylation

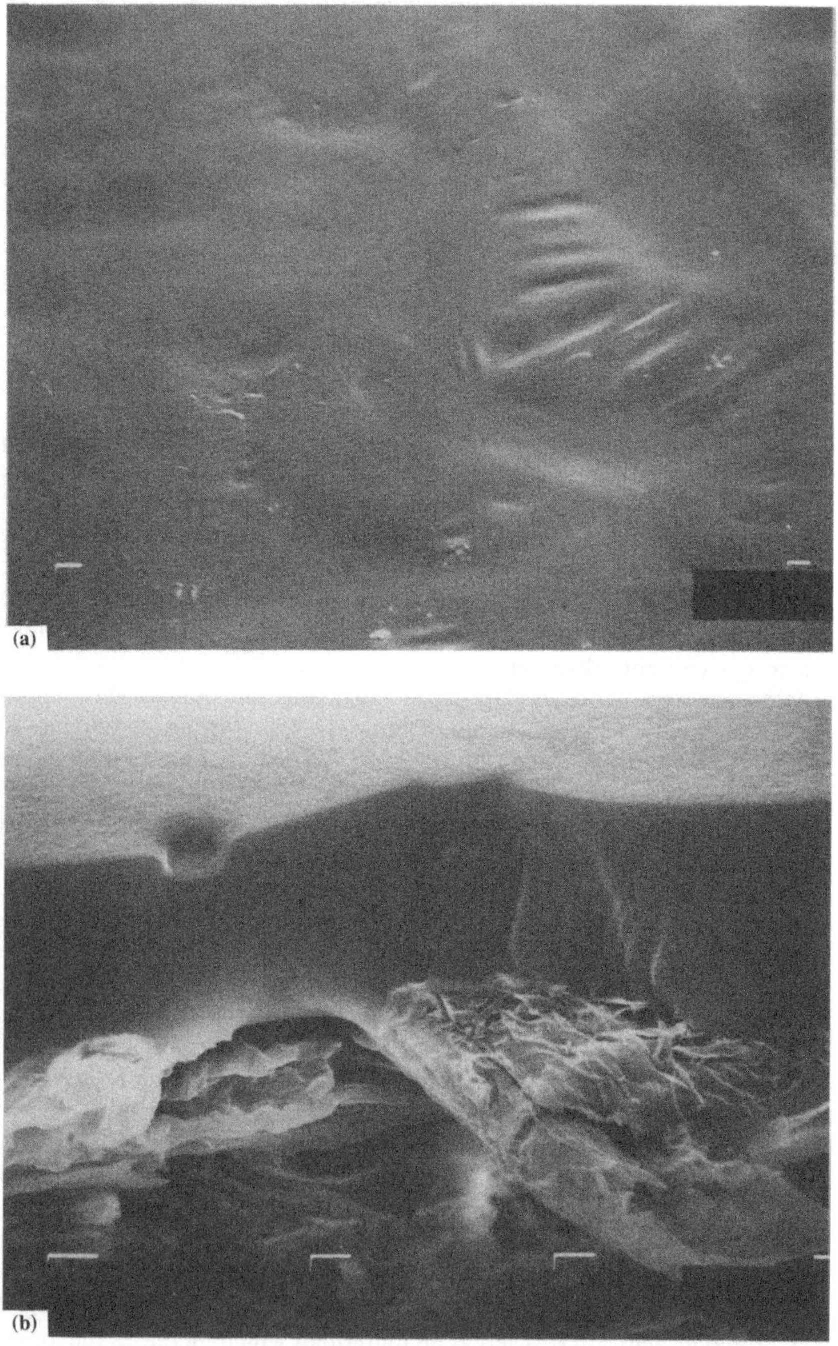

Figure 11.2 Scanning electron microscope pictures of dispersion coating: (a) observed directly from above; and (b) observed from above at a 45° angle.

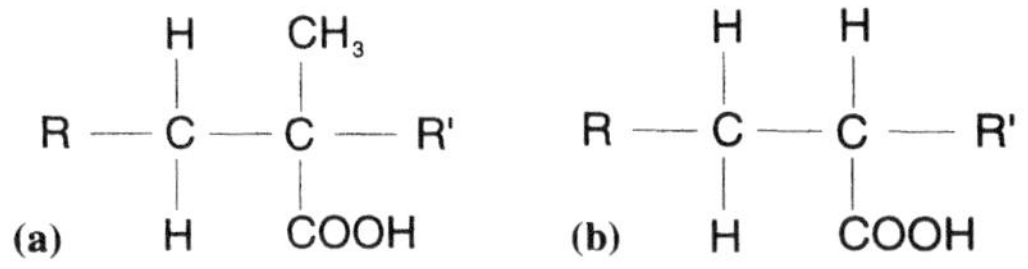

Figure 11.3 The structure of (a) polymethacrylate; and (b) polyacrylate.

has an effect on water retention, viscosity, bonding, barrier properties, etc. As the degree of carboxylation grows the water retention and viscosity both increase exponentially. The water retention factor of dispersion coatings describes the ability of the dispersion to prevent water absorption into the substrate during application and drying. This factor is sometimes ignored or forgotten. It can be used to adjust the degree of penetration of the dispersion into different kinds of substrates. Carboxylation introduces covalent bonds whose bond energies are high: of the order of 100 kcal mol^{-1}. These cause significant changes in chemical, water and heat resistance.

Additives called thickeners are used to adjust the viscosity. Because of this they are sometimes called rheology modifiers. Commonly used thickeners are natural polymers, cellulose derivatives (carboxymethyl cellulose (CMC)), synthetic polymers and starch. On the other hand, the use of thickeners or high micronized wax levels may cause non-Newtonian behaviour (shear-thinning). Plasticizers (for example, dibutyl phthalate) are utilized to decrease the glass transition temperature (T_g) of the polymer by dissolving in the latex particles and swelling them. In general, the dispersion film must be flexible enough to withstand folding and creasing. This ability can be improved by addition of stearates.

Antifoamers are used to prevent foaming before bubble generation; whereas defoamers are used to collapse existing air bubbles. The use of defoamers should be kept at a minimum because certain defoamers have been observed to act as incompatible contaminants and thus to create craters in the coating. Otherwise, surfactants are added to ensure the dissolving of monomers and the initiation of the polymerization reaction. Other commonly used additives are chelating agents, antioxidants, bactericides, biocides, buffers, preservatives, etc. Polymer dispersions can be destroyed, however, by the addition of salt, because salt affects the stability.

Styrene or alkyl acrylate provide hydrophobicity, which is needed when a degree of oil, fat or grease resistance is required. They also allow the formation of gels, infinite networks of particles which may be chemically cross-linked or physically entangled in the case of long chains. This ability to cross-link gives a high level of solvent resistance to the polymer particles because cross-links do not dissolve, though they may swell. Waxes, paraffin, microcrystal or Carnauba wax are all used to change the surface energy and the polarity of the coating. As an example, a 20% addition of wax may decrease the polar component from 30 mJ m^{-2} to 0.1 mJ m^{-2}.

After sedimentation a solid polymer can be sent to landfill. Generally speaking, splashes and small spills of dispersions can be diluted with water and small concentrations disposed of in biologically adapted sewage plants without disturbance to the decomposition activity of the activated sludge. However, it is imperative that the local authorities are consulted to ensure correct waste handling and safety procedures.

11.2 Film formation

11.2.1 Film formation theories

Film formation is one of the most critical phenomena in the field of polymer dispersion coating. Without good film formation it is not possible to achieve the best available barrier levels. Film formation is normally divided into between three to six steps. In principle, the steps are (Figure 11.4):

- shrinkage of interparticle distance;
- coalescence of polymer particles;
- possible further interdiffusion of polymer chains of adjacent particles.

The final properties of the dispersion film are determined during film formation.

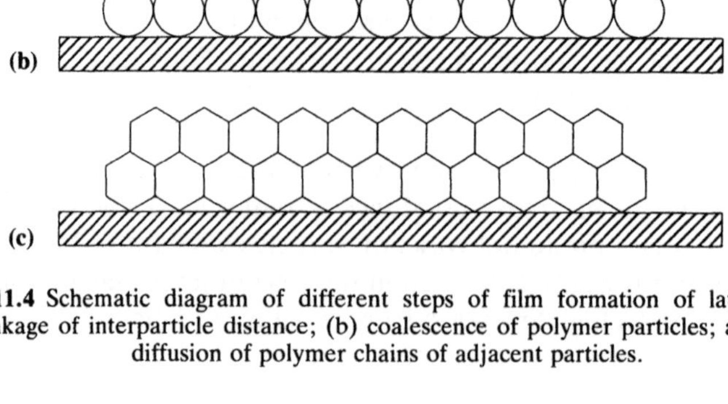

Figure 11.4 Schematic diagram of different steps of film formation of latex particles: (a) shrinkage of interparticle distance; (b) coalescence of polymer particles; and (c) interdiffusion of polymer chains of adjacent particles.

One millilitre of the polymer dispersion contains 10^{13}–10^{16} particles. This means in practice that representation of the film formation phenomenon must be simplified. Typically, this is done by using only two particles in the theoretical models. Various theories and points of view have been put forward in support of the coalescence phenomenon. Sometimes a new theory has supplanted an older theory, whereas another may complement it. Thus, because the film formation phenomenon is not fully explained, the following paragraphs only present according to certain theories possible factors which may produce or prevent coalescence.

It is probable that there are many forces acting simultaneously in opposing directions. Certain of these have an effect over a wide range; and some over only very short ranges, measured in Ångströms. It is, however, important to identify the dominant force at a certain instant and distance.

According to de la Court [1], the following factors may theoretically be important in film formation:

- particle size and particle size distribution – according to Hamaker the attracting force increases as the diameter of the particles increases;
- rheological properties of the polymer particles – the more viscous the particles the more difficult the deformation of those particles;
- water evaporation rate – when the water evaporates too quickly it may not fulfil its task as a medium to bring the latex particles together;
- drying temperature – too low a drying temperature may leave polymer particles so viscous that the attracting forces are unable to deform them;
- chemical composition of the latex – the ratio of co-monomers and the percentage of external plasticizer affects the viscosity of the polymer.

Furthermore, emulsifiers, colloids, catalysts, degree of cross-linking, surfactants, etc. may also influence film formation. The surface energy difference between substrate and dispersion is critical if the dispersions are to spread spontaneously on the substrate. Spreading problems may appear especially when a second coating layer or dye is applied to the dispersion coating.

It is postulated that film formation occurs when the attractive forces exceed the electrostatic repulsive forces. These attractive forces consist of van der Waals forces. On the other hand, the total force, V_{tot}, is the sum of the attractive and repulsive forces, V_{att} and V_{rep}, respectively:

$$V_{tot} = V_{att} + V_{rep} \tag{11.1}$$

This is called the DLVO theory according to Derjaguin–Landau–Verwey–Overbeek [2, 3]. According to DLVO theory, attraction energy has a primary and secondary minimum, and between them there is an energy maximum, the energy barrier. Particles must cross this energy barrier before coalescence can occur.

In the case of polymer dispersion coating, two opposing sets of forces may act simultaneously during the drying process; part of these forces produces coalescence and part resists it. The following list has been extended during development of the theories but Brown was one of the first who originally stated it [4, 5]. Forces which produce coalescence are:

- interfacial tension, F_s;
- capillary forces, F_c;
- van der Waals attraction forces, F_v;
- gravitational force, F_g;
- thermodynamic forces, F_t;
- potential energy required to surpass the maximum repulsive energy, F_w;

where

F_s is a force produced by the polymer–water and polymer–air surface tension, when the particles are in contact;
F_c is the capillary pressure resulting from the presence of a water surface of negative curvature in the interstitial capillary system which is present during the period of water loss;
F_v is the van de Waals attraction between the spheres, that is, the normal attractive forces leading to flocculation or coalescence of emulsions and dispersions;
F_g the gravitational force which leads to settling in a dispersion;
F_t is the force associated with the fact that the polymer chains are highly constrained within the polymer particles – relaxation, mixing and interactions between the macromolecules of adjacent particles are carried out during coalescence;
F_w is the force associated with the fact that the system takes additional energy from the air to reduce interparticle distances – this energy might be used in the first steps of coalescence.

The forces which prevent coalescence are:

- electrostatic or steric repulsion forces between the polymer particles, F_e;
- particle resistance to deformation, F_{ve}.

where

F_e has an effect only before latex flocculation;
F_{ve} is presented as one of the major parameters which control film formation – it depends on the viscoelastic properties of the polymer particles during drying.

When all of these forces are put together the following inequality is formed:

$$F_s + F_c + F_v + F_g + F_t + F_w > F_e + F_{ve} \qquad (11.2)$$

This can be simplified by assuming that some forces are dominant and by taking into account constraints owing to, for example, stability. Moreover, the forces described above do not necessarily all occur at the same time.

The minimum film formation temperature (MFFT) is the lowest temperature at which latex particles can form a solid film after drying. Some researchers have discovered that the MFFT increases when particle size increases. The change in MFFT may be in the order of 10°C if the diameter of the particles varies from 60–460 nm [6]. MFFTs are determined visually by using an aluminium slab which is cooled at one end and heated at the other. The latex is painted in a strip along this slab and the discontinuous point is observed [7]. Another method to characterize the MFFT is to measure the transparency changes of the film.

11.2.2 Rheology

Rheology is the science of flow and deformation, both of which are important in polymer dispersion coating. Dispersions have the behaviour of liquids in the following stages:

- storing;
- agitating;
- coating circulation;
- application;
- metering;
- at the first step of film formation.

After these stages it is more or less a question of the deformation and viscoelasticity of polymer particles. Thus the rheology behaviour is a very important characteristic of barrier dispersions.

Figure 11.5 illustrates some viscosity curves of barrier dispersions. The viscosity values may differ widely but it is very typical for viscosity to decrease quickly when the shear rate is increased. This shear-thinning, pseudoplastic, phenomenon may be due to particle–particle interactions or the non-rigidity of the dispersed phase. In other words, the polymer dispersions do not have Newtonian behaviour at low shear rates but they are quite near to possessing a constant ratio of shear stress to shear rate at higher levels. Barrier dispersions are delivered typically at 500–1000 mPa s (100 rpm, Brookfield) viscosity. Viscosity can be adjusted by water, but if foaming problems occur a one-to-one water–ethanol mix can be used in many cases with good results.

In the case of shear flow, viscosity is defined as shear stress divided by shear rate. Thus the Newtonian liquids have a linear data plot when shear stress is determined as a function of shear rate. Figure 11.6 shows the typical data plot for barrier dispersions. Elevated temperature usually decreases the viscosity, as illustrated in Figures 11.5 and 11.6. The influence of temperature depends on the choice of polymers. The effect of

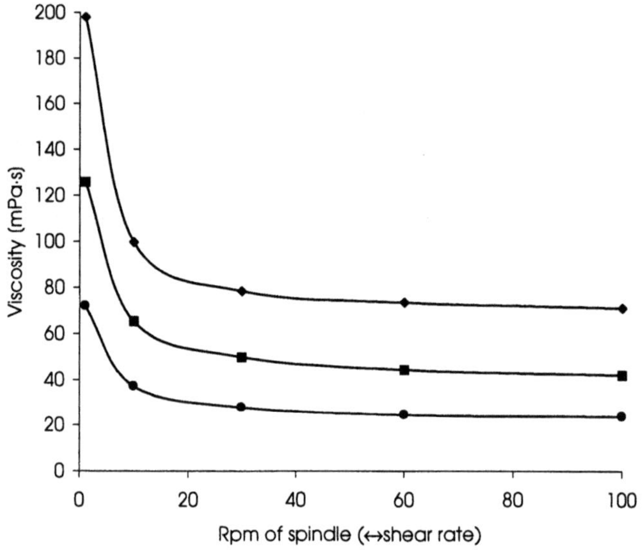

Figure 11.5 Typical viscosity behaviour of styrene–acrylate copolymer barrier latex, measured by Brookfield viscometer. Temperature: ◆ = 14°C; ■ = 34°C; ● = 53°C.

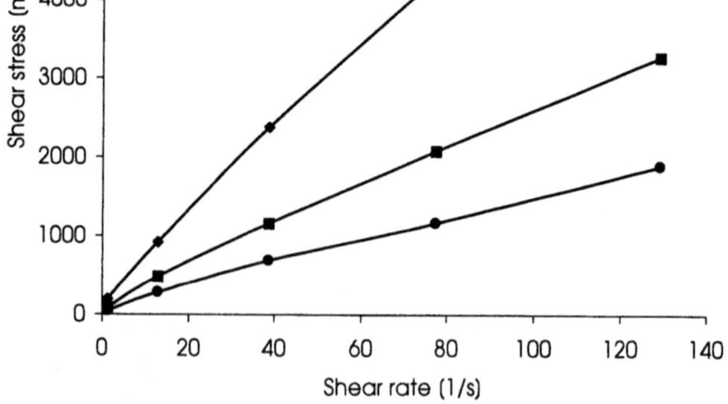

Figure 11.6 Shear stress as a function of shear rate for styrene–acrylate copolymer barrier latex. Temperature: ◆ = 16°C; ■ = 39°C; ● = 56°C.

temperature may be important in practice if the environment of the coater is clearly divergent from normal room temperature.

After application and metering, excess water is evaporated from the coating, and the polymer parties start to coalesce, following which the viscoelasticity of the polymer particles plays a significant role because the particles must be able to lose their original shape and identity to form a uniform coating layer.

If the drying temperature and time are not suitable for the dispersion it is possible that particles do not form as tight and uniform a structure as possible. Figure 11.7 shows a sample of a badly dried dispersion coating on board. Polymer particles have not totally lost their spherical shape because of the temperature-dependent viscoelestic properties of the polymer. This has had a bad influence on the barrier level and mechanical properties. The glass transition temperature (T_g) is widely used to describe the hardness or softness of the polymer. It can be modified by, for example, changing the polymer ratio in the case of a copolymer or by addition of a plasticizer. Viscoelasticity can be determined by dynamic mechanical thermal analysis (DMTA) equipment [8].

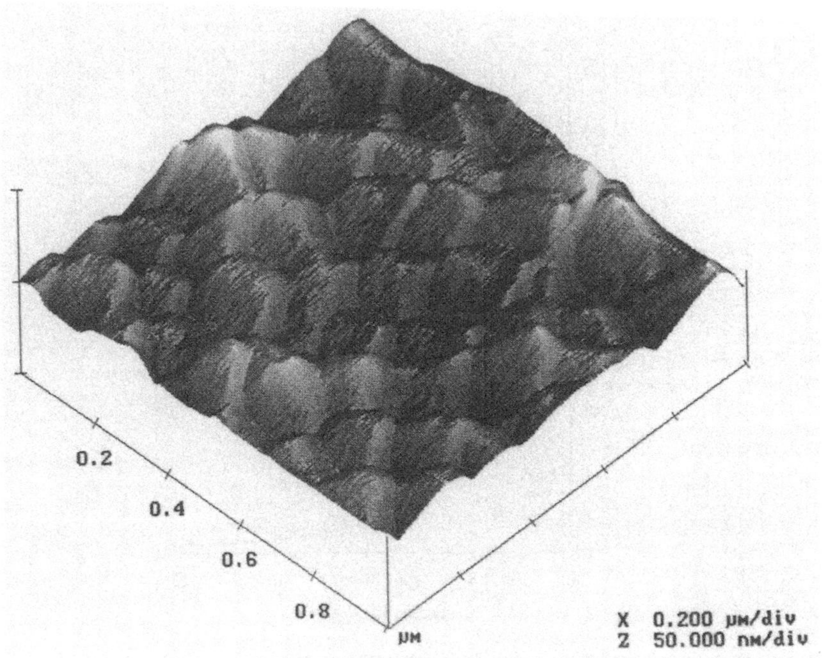

Figure 11.7 Image of badly dried dispersion film.

11.3 Commonly measured film properties

11.3.1 Water vapour transmission rate

Water vapour resistance (measured by the water vapour transmission rate (WVTR)) is the most frequently required barrier property. It is also sometimes called moisture vapour transmission rate (MVTR). WVTR is normally measured by the cup method or by special WVTR equipment. These methods determine somewhat different phenomena and that is the reason why there is a difference in the values obtained by the two methods. This difference may be as much as 100%. The unit of WVTR is g m^{-2} per 24 h. Typically used measuring conditions are:

- 50% relative humidity at 23°C;
- 75% relative humidity at 25°C;
- 90% relative humidity at 38°C.

The last of these is the so-called tropical environment, though slightly different conditions are sometimes called by this name as well. Because the measuring condition has a remarkable effect on the values achieved it is always important to check the conditions used.

WVTR values vary widely, depending on the substrate, coating technique and dispersion utilized. Figure 11.8 illustrates the typical water

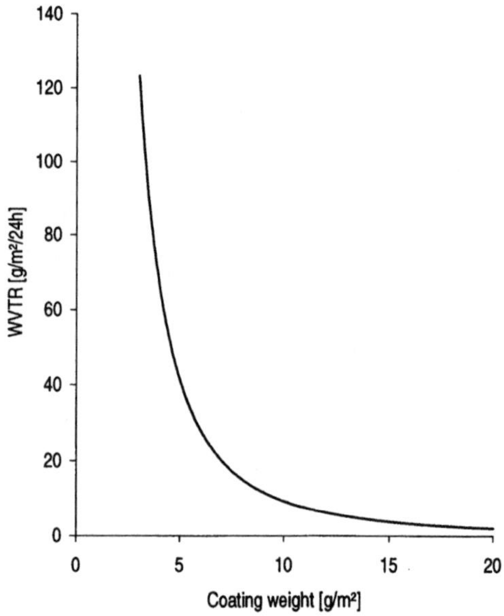

Figure 11.8 Water vapour transition rate (WVTR) values of a typical polymer dispersion coating: single coating.

vapour transmission behaviour of a barrier dispersion at different coating weights. If the coating weight is not too low (a minimum of 3–5 g m^{-2}), better values are achieved by dividing the coating into two layers. However, double coating requires two coating units or two runs.

A coarse estimation is that in WVTR values a 10 g m^{-2} dispersion film is equal to the same grammage polyethylene film. This estimation assumes that the dried dispersion film is pinhole free, because pinholes can decrease the WVTR value considerably.

11.3.2 Cobb value

Barrier requirements should be adjusted to fulfil the demands of a particular application. One of the barrier demands is water absorption, which is characterized by the Cobb value. The Cobb value describes the amount of water absorbed into the sample within a certain time period. In the case of barrier dispersion coatings, useful measuring times are 300 s and 1800 s, depending on the laboratory capacity and the coating weight. Even longer times are used for excellent coatings. However, measuring times can be shortened by utilizing increased pressure.

The Cobb value decreases (water resistance increases) sharply after a certain coating weight is reached (Figure 11.9). This level is, in practice, the minimum sensitive coating weight. It depends on the film formation

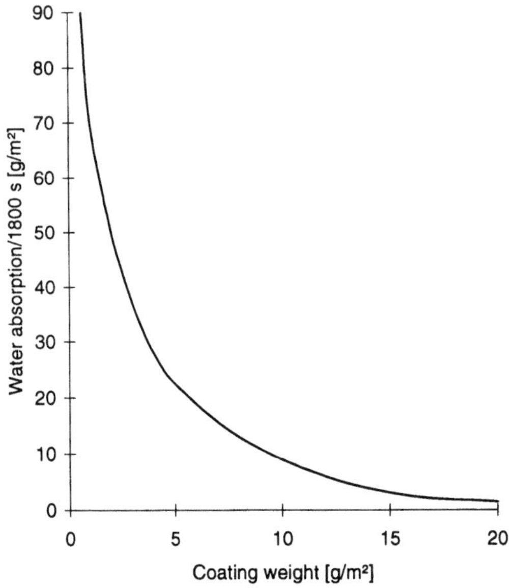

Figure 11.9 Water absorption as a function of coating weight.

ability of the dispersion and on the surface properties of the substrate as well as on the coating technique. For pigment-coated grades the minimum sensible coating weight is typically 3–5 g m^{-2} and 2–3 g m^{-2} more for uncoated substrates.

It may seem self-evident that a dispersion film that possesses good water vapour characteristics will also provide good water absorption values, but this is not necessarily true. A typical example of this is unfilled styrene–butadiene latex. It may give excellent WVTR values but a relatively high water absorption rate. Thus it is important to know the critical demands and to separate them from desirable requirements. Also, it is important to determine the minimum barrier level in order to avoid overcoating; in some cases the present barrier structures are overdesigned.

Water penetration through the polymer film may change many film properties. This is because of the fact that when a polymer absorbs water its T_g value decreases because of the plasticization effect of water. Also, if water penetrates through the coating into the substrate the fibres start to swell and this may cause local damage to the coating and lead to increased penetration levels. Barrier dispersion films are not typically highly flexible, which may cause problems during creasing because it is possible that the edges do not sustain the operation and thus the barrier properties are lost. This is not such a problem if the packaged contents consist of particles; however, powders and liquids are more sensitive.

11.3.3 Other properties

Even a short contact with greases, fats or oils can easily leave stains on fibre-based packages. These problems can be solved by using barrier dispersions. Certain dispersions are especially designed to provide grease and oil resistance. These properties are achieved by, for example, a high degree of cross-linking. Oil, fat and grease resistance is commonly tested with:

- dyed turpentine;
- the Kit test;
- the chocolate test;
- the oil absorption test.

Moreover, barrier dispersions can be used against the following high molecular weight materials:

- mineral oil;
- paraffin oil;
- peanut oil;
- olive oil;

- mineral greases;
- animal fats;
- fish oil;
- vegetable oil;
- molasses.

Odour and taste barriers are needed to protect the contents from unwanted odours or to protect the environment from the odour of the contents. Polymer dispersion can be used for this purpose if the film is dried perfectly. However, some dispersions may themselves produce strange flavours if they contain, for example, monomer residues. It can be said that dispersion films typically cannot in practice decrease the odour of the substrate itself but do not necessarily increase it either. This underlines the importance of the substrate. Thus the best possible odour properties are typically achieved by solid boards with well-dried barrier dispersions.

Odours are typically tested by human nose because it is one of the most sensitive testing instruments available. However, the results depend on the ability of the test panel and they may vary from one panel to another. The scale of the odour test is commonly from 0–3 or from 0–5, where 0 represents odourless and 3 or 5 a strong, offensive odour. The level for solid board is typically 0.5, and after polymer dispersion coating the value is typically from 0.5–1. However, if the dispersion coating is badly dried the value may easily be 2.5 when the maximum value is 3.

Fillers can be added to polymer dispersions in order to achieve a certain ultraviolet light barrier level. These properties are needed if the contents are greasy or contain other light-sensitive materials. Commonly utilized fillers are the same as those used in coating colours: titanium dioxide, talc, calcium carbonate, etc. The appearance can also be modified by addition of fillers or dyes.

It is always the target to achieve pinhole-free coating. Pinholes are small holes through, or nearly through, the film. Pinholes are usually created by collapsed air bubbles or *z*-directionally orientated fibres. The presence of pinholes is revealed with so-called pinhole test liquids. These liquids may consist of water, surface active agents and dye or they may just be low surface energy liquids with a small amount of dye added. Different kinds of dispersion films require specially designed liquids to make sure that the film is not damaged during testing, yet ensure that all pinholes are found. If specially designed liquids are not available, colour plus turpentine or spirit can in many cases be used, as well as washing detergent and colour-containing water.

Many packaging applications require a gas barrier from the coating. In future these properties may be achievable with polymer dispersion films. Oxygen and carbon dioxide barriers can be made with copolymers of highly polar polymers and non-polar polymers.

11.4 Heat sealing, blocking and gluing

The surfaces of the product must be able to be joined together in order to form the final structure. Sometimes the joining is done mechanically (for example, in many corrugated board applications) but in many cases it entails heat sealing or gluing. However, the dispersion coated surface should not block or stick on the adjacent surface under any circumstances. Unfortunately, good heat sealing properties and antiblocking properties seem to be, at least partly, contradictory. This is the case for soft polymers which typically have good sealing properties but which may also block quite easily.

Commonly used heat sealing methods are bars, hot-air, ultrasound and impulse sealing. Sealing temperature, hold time and pressure are different for these various sealing methods. That is why some dispersions are easily sealed with a bar sealer but may have sealing problems with, for example, a hot-air sealer. Additives, waxes, carboxylation degree and fillers each have a remarkable effect on sealing properties. The effect may be so dramatic that the same dispersion without wax is easily heat sealed and after wax addition totally loses the property. Depending on the utilized polymer, the dispersion may be more sensitive to pressure or to temperature.

Blocking tendency is measured either by using a short-term but high-impulse force or by a low-intensity, long-term force. Temperature, moisture content, pressure and exposure time all have an effect on blocking. Blocking may occur inside the roll when rewinding is difficult or it may occur inside stacks of blanks. Typically, blocking happens directly after coating, but unfortunately it may still happen weeks after the coating process. This is the case when the final products are, for instance, delivered in a hot cargo hold of a ship. Thus the blocking occurs at the customer destination although the product was in good condition before shipping.

Some dispersions do not block, generally speaking, under any circumstances, but neither are they then heat sealable. Many heat sealable barrier dispersions are block resistant up to 50°C, according to the manufacturers. Thus the layout of the machine must allow sufficient cooling time to the coated substrate or additional cooling devices must be utilized. Cooling raises financial and space questions on wide and fast paper or board machines. To avoid the necessity for additional cooling, the drying method should be such that the wet coating absorbs the drying energy with minimum warming of the substrate. A high solids content of the dispersion minimizes its drying requirement, thus preventing blocking.

The blocking tendency is difficult to measure on-line if the critical temperature and/or moisture content is unknown. In the laboratory, temperature, pressure and time factors can be easily controlled simply by adjusting the press inside an oven. Moreover, the relative moisture content can be controlled by placing the oven inside an environment-controlled room, where possible. Certain fillers can be used to eliminate blocking but can also have an adverse effect on the heat sealing properties.

Barrier dispersion films should be gluable with common starch or vinyl-acetate-based dispersion glues. Unfortunately, this is not possible in many cases. This is because of the nature of the gluing process. Glues contain water, which has to be able to penetrate through one or both sides of the substrate. In the case of a barrier coated substrate this ability is limited when the adhesion between glue and substrate remains weak. However, hot melt glues can normally be used to join dispersion coated surfaces.

11.5 Products

Polymer dispersion coatings can be used in nearly all packaging applications where polyethylene (PE) has so far been dominant. In addition, there are several applications where PE cannot be used but polymer dispersion coatings can be utilized without problems. Until now the commercial products have been produced mainly by separate coating lines or presses. Polymer dispersion coatings are used on different kinds of wrappings, sacks, disposables, frozen and chilled food cartons and corrugated boards. Most of the dispersions on the market can be used in food applications, but there are special coatings, for example, with a high detergent resistance in order to be used on detergent packages. One quite interesting possible target for dispersion coatings is to use them as precoating before, for example, polyethylene film. This kind of structure allows reduction of extrusion film without losing the barrier properties. On the other hand, dispersion and extrusion films can give different kinds of barrier properties to complement each other. Typical products require from the coating not only the barrier properties but sealing or gluing ability and good recyclability.

Polymer dispersion coatings have only been in use for a short time and it is evident that they have not yet reached the maximum possible barrier levels. In practice this means that the initial applications, such as for frozen vegetable boxes and ice-cream wrappings, have been less demanding. Ice-creams are easy because they need only a minimum barrier coating in storing and have a short active life after melting. Typical polymer dispersion coatings are compostable, which has led them to be used with disposables where the active life of the ingredients is also short. Typical dispersion coated disposables are cups, plates and trays. After use the barrier dispersion coated disposables can be composted or, if left in nature, they will visually disappear within one or two months. For these same reasons barrier dispersions are also used on sacks.

Recently, the barrier levels of dispersions have increased, allowing more demanding applications with reasonably low coating weights. One remarkable sector where barrier dispersion can be utilized is for industrial wrappings and headers of paper rolls. The volume of these products is so high that the required properties must be achieved with minimal costs. The WVTR protection of ream wrappings of copy paper are achievable by dispersion coating. Non-food applications do not in many cases need to be taste or odourless, which makes the coating process less demanding.

The above-mentioned applications need mostly water resistance or water vapour resistance. Bakery and confectionery packages, as well as candy wrappings and boxes, need grease and oil barriers. Dispersion coatings are very suitable for these applications because the dispersion films normally survive either microwave or standard ovens. This property must, of course, like other properties, always be checked beforehand by the manufacturer of the dispersion. Other applications are, for example, pet food packages, electronic packages, hygienic and garden product packages, wallpaper base, label paper, other industrial wrappings, etc. Dispersion coatings can be used for the interior coatings packages for dry, solid and powder goods such as chocolate, cookies, cocoa powder, etc., so that inner pouches may be omitted.

11.6 Disposal of dispersion coated papers

11.6.1 Recycling

Polymer dispersion coatings are as a rule repulpable. Repulpability of dispersion coatings allows the fibre material of barrier coated paper and board products to be recycled. It is an essential property in order to lower the load to landfills and incineration plants. Thus, products coated with polymer dispersions have acquired a 'green' image. The repulpability quality makes on-line coating sensible in that it allows the re-use of coated broke without special treatment.

Repulpability properties depend on:

- coating thickness – thick coatings are more difficult to repulp and the process takes more time than in the case of the thin coated or uncoated substrate;
- existence of precoating – suitable precoating prevents the barrier dispersion penetrating inside the substrate, thus shortening the required pulping time;
- hydrophobicity of the coating – hydrophobicity increases the wetting time of the fibres, the swelling breaks the coating, furthering the wetting and the swelling progress;
- the coalescence level of the polymer particles – if polymer particles are just packed one on top of another they can be easily separated, if instead

polymer particles are coalesced perfectly and polymer chains of adjacent particles are interdiffused the separation process is slower.

Laboratory repulpability tests can be done either by a pilot pulper or by a laboratory disintegrator. In the pulper the separation force is applied mainly by friction and shear forces from other fibres, but in a disintegrator the force is applied by the stirring device. This inequality is a result of the consistency difference and speed difference of the stirring rotor. In real pulping processes there are nearly always some chemicals involved. This chemical addition makes nearly all pulping processes different.

It is evident that the repulpability itself does not guarantee that the dispersion coated materials can be re-used on paper or board machines. With these machines wet-end chemistry is in general sensitive to colloidal particles and is not allowed to be disturbed because that has a direct reflection on the quality of the product. The chemical balance of the machine circulation is not the only variable where the use of polymer-dispersion-containing broke may have an effect. Although the polymer particles are separated during the pulping process they may coagulate afterwards to form sticky particles. These particles may cause dirtiness on the wires in paper or board machines and missing dots in printing.

Because of the wide interest in the application of barrier dispersion coatings on board or paper machines experiments have been run on production machines. From the results of these runs, recycled pulp containing dispersion coatings can be used without problems. In fact, some of these tests have been done without telling the employees. The results have been that the machine operators have not noticed the tests, indicating that the polymer particles can pass through the wet end and be incorporated in new paper or board. But, the operators always have to ascertain the suitability of pulp containing dispersion coatings from case to case.

11.6.2 Composting

Compostability is, in addition to repulpability, an advantage of barrier dispersion coatings compared with polyethylene films. Here the compostability is understood to mean the 'disappearance' of the material in the compost environment by visual observation. Compostability allows the use of barrier coating in, for example, bio-waste pouches.

Sausage casings have been used as a positive control in tests where styrene–butadiene (SB)-latex film and SB-latex coated kraft paper are composted. Figure 11.10 shows that latex film starts to compost immediately and dispersion-coated kraft paper after 15 days, like the sausage casing. After 35 days the sausage casing is almost completely composted, whereas the dispersion film and the dispersion-coated paper seem to need 5–10 days

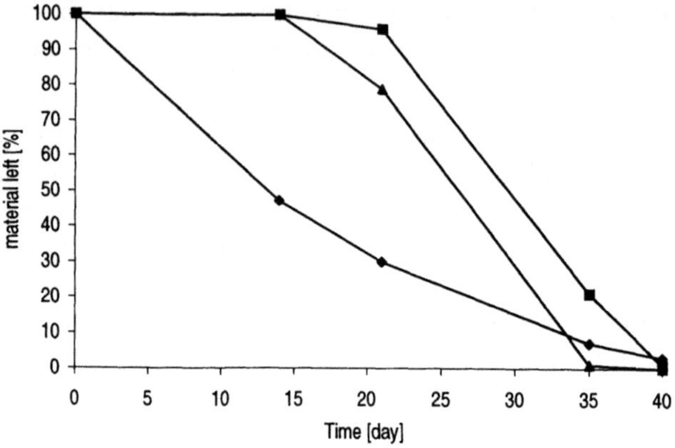

Figure 11.10 Degradability of styrene–butadiene-latex in a compost environment, by visual assessment. ◆ = SB-latex film; ■ = SB-latex-coated kraft paper; ▲ = sausage casing. Reproduced with permission from an unpublished research report by Kari Nurmi, R&D Manager, Oy Raisio Chemicals Ab, 1995.

more. These results are based on visual assessment. Many other composting tests have shown similar results. In general, the degradation time depends on environmental factors (humidity, pH, temperature, other compost material, oxygen and carbon dioxide content, etc.) of the compost.

References

1. de la Court, F.H. (1970) Environmental conditions and the Films Formation of latices. 10th FATIPEC Congress, 293–7.
2. Derjaguin, B.V. and Landau, L.D. (1941) *Acta Physicochemica*, **14**.
3. Verwey, E.J.W. and Overbeek, J.T.G. (1948) *Theory of the Stability of Lyophobic Colloids*, Elsevier, Amsterdam.
4. Brown, G.L. (1956) Formation of films from polymer dispersions. *Journal of Polymer Science*, **12**, 423–34.
5. Leonardo, R.-G. (1990) Film formation: aspects on the coalescence of latex particles. *Makromol. Chem., Macromol. Symp.*, **35/36**, 389–404.
6. Jensen, D.P. and Morgan, L.W. (1991) Particle size as it relates to the minimum film formation temperature of latices. *Journal of Applied Polymer Science*, **42**, 2845–9.
7. Protzman, T.F. and Brown, G.L. (1960) An apparatus for the determination of the minimum film temperature of polymer emulsions. *Journal of Applied Polymer Science*, **4** (10), 81–5.
8. Rantanen, T. and Kimpimäki, T. (1995) Thermal and mechanical characterization of biodegradable polymer coatings. Annual Meeting of the Nordic Rheology Society, Helsinki, 7–9 June, 1995.

12 Recycling surface treated paper products

D. GUEST

12.1 Introduction

Recycling once-used paper is a well-established process within the international paper and board industry, accounting for approximately 35% of the raw material used. Its importance as a raw material is growing with predictions [1] that it will exceed 45% of the raw materials used by the global paper and board industry by 2010. Figure 12.1 illustrates this.

No systematic study has been made of the influence of surface treatment and treatment chemicals on recycling. Some effects have been documented, particularly those that are detrimental. The key issue to many is the extent to which surface treatment chemicals may restrict the ability to increase recycling rates, particularly with the trend to surface treat more grades for functional and process reasons. This must not be taken out of context, as other trends, particularly to lighter weight papers, also have an impact that may be greater.

Historically, the reason for the success of waste paper as a major papermaking raw material is economic. The total cost of purchasing waste-paper raw materials and processing them into marketable products is less than the competitive alternatives provided by virgin pulps. Because the recycling of paper is a mature industry a broad balance exists between the availability of waste-paper grades with the properties required for the manufacturing process to convert them efficiently into products. Factors that contribute to this balance include the collection cost and waste-paper price relative to pulp. Legislation and subsidies that tend to favour waste paper are also important.

As with any other product, there are periodic disturbances in supply and demand that increase or decrease the competitiveness of waste paper. Along with many papermaking raw materials, prices vary widely with the point in the economic cycle. Although there is a relationship between waste-paper and pulp prices there are some fundamental differences. It is possible for pulp mills to take downtime and either stockpile chips or delay harvesting. This option is not readily available for waste-paper users as there are logistic problems of storage. Unlike wood or chips the papermaking potential of waste paper falls significantly when stored.

The recovery rate of waste paper expressed as the proportion of paper

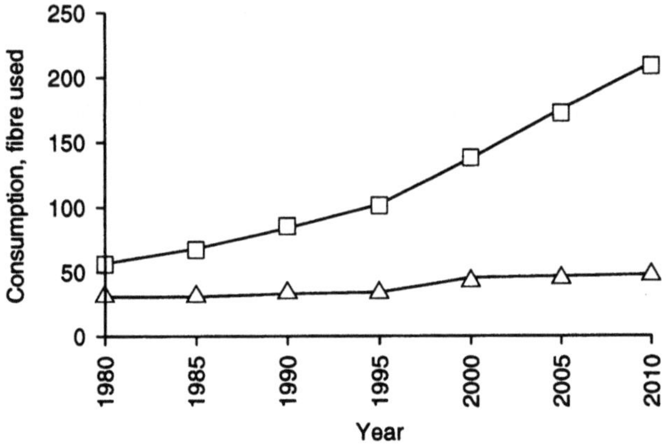

Figure 12.1 World consumption of waste paper. □ = consumption (mega tonnes per annum); △ = percentage fibre used.

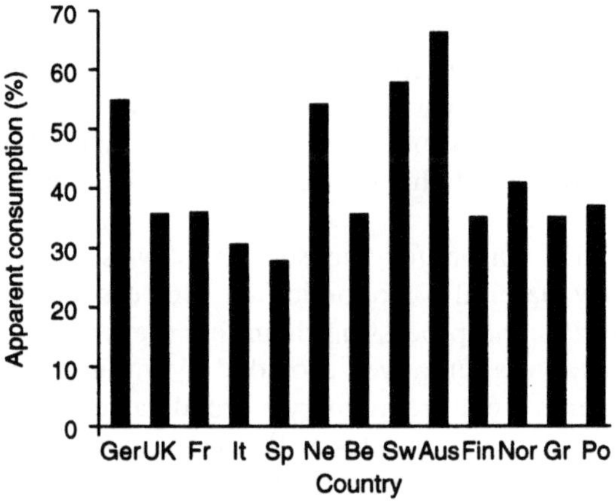

Figure 12.2 Paper recovery rates for Europe. Ger = Germany; UK = United Kingdom; Fr = France; It = Italy; Sp = Spain; Ne = the Netherlands; Be = Belgium; Sw = Sweden; Aus = Austria; Fin = Finland; Nor = Norway; Gr = Greece; Po = Poland.

collected for use as an industrial raw material to the total amount consumed varies considerably from one country to another. Figure 12.2 illustrates the recovery rates achieved in 1993 or 1994 for European countries, in order of their apparent consumption [2]. Clearly there are large differences within the developed economies, with the Austrians, Dutch, Germans and Swedes in the lead. The United Kingdom and southern European countries with lower environmental pressures lag. Recovery rates also reflect the balance between the production of various grades and the investment made by the country in recycling capacity and the collection infrastructure. Population densities also have an influence and partly explain the low recovery rates in Finland and Norway, even where there are considerable environmental pressures.

When examining anything to do with paper and board it is easy to forget that they are substrates or structural materials that find a use within a number of quite distinct value networks. Broadly, their use is in the packaging, information, communication, security, hygiene and construction networks. The functional requirements are very different for each application. It is therefore not surprising that the market penetration for waste paper as a raw material is vastly different for each paper product. In packaging grades it provides 80% of the raw material in Europe, whereas for high-priced coated papers its utilization rate is less than 10%. Figure 12.3 illustrates typical utilization rates for waste paper in a number of paper products [1].

The historic concentration of waste-paper consumption in lower value added packing grades is changing slowly for market and environmental reasons. It illustrates, however, the relative difficulty and costs of processing waste paper, including that which has been surface treated.

Packaging grades for the distribution of goods exploit the structural properties of paper and board. The visual appearance is less critical and the recycling plant installed is comparatively simple. In contrast, for high added value publications grade products visual appearance is critical to function and the perception of consumer value. A high brightness is essential for marketing reasons and the paper must be speck-free. Cartonboards have surface property requirements similar to those of

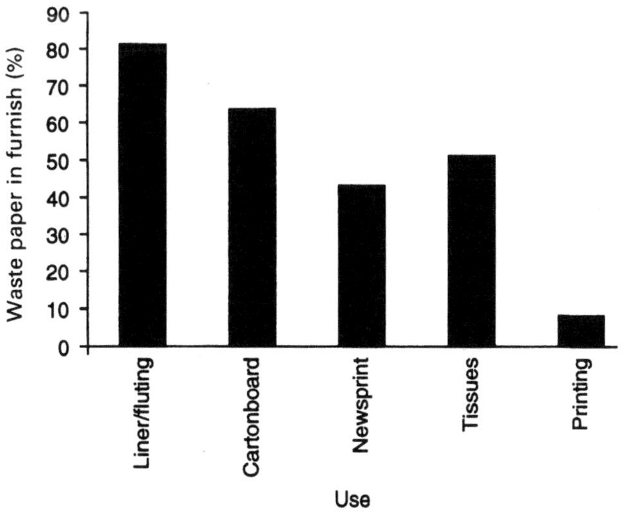

Figure 12.3 Waste-paper utilization in different product groups.

publication grades, whilst the appearance of the inner layers is not critical. More importantly there must not be migration of materials from the board into foodstuffs it may contain. Tissues need to be bright and clean for marketing reasons and free from mineral coating or fillers for functional reasons.

12.1.1 Surface treated paper in context

The properties of waste paper needed for various grades of paper and board are quite different. Generally, the more demanding the application the greater the cleanliness and homogeneity required.

Paper is surface treated within the mill, usually with starch or a complex coating slip containing minerals, synthetic latices, starches and a range of other additives. Wastes from the process, either off-specification product or trim, are recycled internally within the mill. This is the most homogeneous waste and finds application in a product of similar added value to itself.

Most paper leaving a mill is converted in some way, for example, by printing or box making, to enable it to perform a useful function. The converting operation is seldom 100% efficient, with off-specification product and trim resulting. This industrial or preconsumer waste is between 20% and 30% of consumption. With returns of unsold publications of between 20% and 50% it is for many countries the basis on which their recycling industry functions. Such waste arises in bulk on single sites and is relatively homogeneous. The waste merchant contracts to remove the waste from converters' premises and adds value by sorting into grades suitable for the production processes installed within mills.

Waste paper is an internationally traded commodity using a complex set of grades. Some relate to the fibres used in the manufacture of the original product whereas others are far less specific, describing the function performed by the paper or its collection source. The value depends on the supply and demand situation and the range of impurities found in the waste. Certain converting processes such as lamination with metal foils and plastic, as in liquid packaging board, and silicone coating for self-adhesive labels reduce its value significantly. This is a consequence of the difficulty of recycling such impurities in all but the most sophisticated and dedicated plants. Often this type of waste has a negative value and is disposed of by landfilling or incineration.

Much of the new capacity based on waste paper uses post-consumer sources. Here it is not so much the individual surface treatment chemicals that are the problem but the wide range found. The driving forces for the use of post-consumer waste have been both market and, more importantly, environmentally related. The environmental concerns are a response to the problems met in many countries over the disposal of solid waste as landfill sites become exhausted. Paper and board products account for approximately 35% by weight of landfill materials originating from households. Not surprisingly, they have come under close scrutiny by the environmental movement and legislators.

12.1.2 Solid-waste management policies

Environmentalists and politicians agree on a hierarchical approach to solid-waste management. Prevention is at the top of the hierarchy, followed by re-use, recycling, incineration with energy recovery and incineration, with landfill firmly at the bottom. The manufacture of lower added value products and composting lie somewhere between incineration without energy recovery and recycling. Some argue that landfill is the only true waste management option in the hierarchy because the others contribute something to societal well-being and add value to the waste. A balance must be struck between the amounts of waste paper disposed of by each route to minimize the total economic and environmental costs. Surface treatment with chemicals can change the environmental costs by, for example, reducing the weight of material needed to perform a function and improving performance when recycled.

The manufacturer of a consumer durable can reduce the solid-waste burden by better design to extend their products' lifetimes. Paper is a short-lived consumable – in the case of newsprint it can be made on Monday and be back in the de-inking plant by the end of the week. However, paper manufacturers can prevent waste by improving control of the process to reduce variability. This allows a reduction in basis weight, thereby saving material. A number of developments in machine design and sensors drive this. An example from the papermachine is cross direction (CD) basis weight control. The replacement of slice screws to adjust stock flow in headboxes by dilution control to the feeder pipes lessens CD variability. The better the consistency at this point the less surface applied chemicals are needed to perform a given task. It also speeds up grade changes thereby reducing process waste. On-line control of coat weight is becoming standard for modern pigment coating installations and is being retrofitted elsewhere. These will develop further as the control companies start to use open systems to incorporate the best available sensors and control algorithms.

The idea of saving waste at source is a potential threat to the paper industry. Offices with recycling schemes become aware of their paper consumption and cost. The much quoted 'statistic' that each piece of paper entering an office generates 14 other sheets has an element of truth. However, as the number of service sector employees contracts, so will paper consumption. Companies that use considerable quantities of paper are already starting to replace many of the storage and communication functions of paper by using the power of electronic office equipment.

Paper and board products fulfil a number of functions each with a different lifespan that has an influence on our ability to re-use them. Some products, such as bank notes and re-addressable envelopes, are designed for re-use. Grades manufactured for building applications may have a relatively long lifespan, their replacement depending on consumer response and attitude rather than failure in service. It is very difficult to envisage how these products could be re-used. Tissue products cannot be re-used after they have fulfilled their primary purpose nor can printing papers that provide permanent records, either for business transactions or in books.

The redesign of some transit packaging grades could increase their capability for re-use and there are one or two novel designs originating from Germany. In this application higher basis weights are needed for structural reasons as the pack has to perform through a number of distribution cycles. The environmental balance between the extra resources consumed in manufacture and transporting the additional weight of packaging for a multitrip pack and a conventional paper product is complex. However, it must favour re-usable packaging.

An opportunity for surface treated papers lies in reversing the printing process to recreate plain paper. For this to work, digital printing technology could activate a thermal or light-sensitive coating to produce the image. To re-use the paper a digital scanner would identify where deactivation is needed to remove the printed image creating a new surface, possibly still containing logos and sales messages. It remains to be seen whether there is the economic incentive to develop something along these lines. Some technology already exists as evidenced by security papers that lose their image when photocopied.

12.1.3 Design for recycling

The economic and environmental balance between the various waste management options can be altered by giving designers a broader brief to look at the recyclability aspect of products. So far attempts to design products that are easier to recycle concentrate on improving adhesive and ink formulations, but these developments are always in the context of the current installed technologies for processing waste paper. This again is natural to protect the installed base but may be a suboptimal approach in the long term. Essentially, the motto 'all for one and one for all' applies. An example of this is Mead Corporation's development of a novel repulpable wet strength additive [3]. However, there is little control of the post-consumer waste. The commercial and environmental benefits of this new approach can only be realized by ensuring that all wet strength papers and boards are manufactured by the same chemistry. This in part explains Mead's willingness to license their technology to competitors.

The overall approach to designing recyclability into products has implications within the surface applied chemicals sector. When developing new applications, the implications for the economics of recycling becomes more important . Modelling the fate of the new chemical as it progresses through the waste recovery and recycling system is then essential. Some companies are already looking at chemicals that convey advantages when recycled to differentiate them from competitors. In this context cationic starch is being promoted on the strength benefits it provides to furnishes when recycled. Other chemicals that remain adsorbed to furnishes could be designed with extra functionality to enable them to perform a useful task within waste-paper treatment plants or papermachines.

In many ways recycling is the highest waste management option open to the paper industry. This does not necessarily mean that the paper industry should recycle all the paper recovered. The paper industry, like many others, adds value to its products by introducing attributes and features such as colour, coatings and laminates. As it proceeds down the value network others enhance its worth by converting processes that add ink, adhesives and other chemicals. When we consider the important features of a raw material, its consistency and homogeneity are always high up the list. Added to this, we are looking for freedom from impurities and the ability to process it with high yields. The processes of adding value to paper and paper products make it more heterogeneous and lower its worth as a raw material for paper manufacturing.

The barrier to recycling is largely economic and not technical. It depends on the yield, the energy needed to convert a heterogeneous raw material into one suitable for papermaking and the value of the paper produced. The recycling of fibres is a multi-dimensional problem and there can be no fixed answer to the question of optimum recycling rates. The danger is to assume that a single aspect of paper manufacture, conversion or use limits its recyclability. In practice, surface applied chemicals have an impact that ranges from major to insignificant, though they are difficult to quantify. This must not be taken out of context with other materials that have different impacts. Recycling only becomes uneconomic or an environmentally less-preferred option when the sum of the impacts from all materials becomes too high.

This type of approach has catalysed a growing realization that recycling forms part of a broader waste management strategy. This permits incineration with energy generation to utilize the calorific value of that waste paper which is uneconomic or environmentally harmful to recycle, for example, for process reasons, because of low population density or simply to bleed off some of the waste to maintain the overall properties of the raw materials.

There comes a point, as the level of recycling rises, where waste paper becomes uneconomic. This is primarily a result of the properties of the major material used, cellulose fibre and the property changes that occur on recycling. These are influenced by the process of surface applying chemicals.

12.1.4 The influence of fibres on paper recycling

Fibres used in the manufacture of paper originate from a variety of plant sources. In the developing world cereal and other plant residues are important sources of papermaking fibre. However, in the global context they account for less than 5% of the material used. Wood pulp is the major papermaking raw material. This wood originates in managed forests or plantations grown for the specific purpose of cropping. In many cases the pulp industry processes the residues from other wood products, for example, sawmill waste. Common species used are pine, spruce, birch and eucalyptus, though many others find commercial application.

A number of pulping processes convert wood into a usable pulp. These fall into two main categories, chemical and mechanical, though there are hybrid processes. Essentially, in chemical pulping the wood is subjected to an acid or alkaline process by using sulphur-compound-based catalysts that attack the lignin and lignin–hemicellulose bonds. The lignin is solubilized and the fibres released with an overall yield of approximately 50%. In contrast, mechanical pulping relies on mechanical energy physically to separate the fibres from each other. Yields are in the 93%–98% range. Mechanical pulp contains all the original wood polymers and this has a major influence on its stability to ultraviolet light and gives rise to the characteristic yellowing. This has an influence on its recyclability.

The process of drying pulp fibres causes irreversible changes in their properties and structure [4]. The extent to which this occurs depends on the wood species and the pulping and bleaching methods used. The papermaking process also exerts a major influence. The use of chemicals can alter the surface of the fibre, whereas refining treatment and pressing both influence its structure. However, these are secondary effects, with the major influence on recycling properties being the restraint during drying, the drying intensity and the moisture after paper drying. The effects are noticeable at 20% moisture and become more critical as the fibres are dried to below 10%. This is important because many paper grades now use film applicators and size presses to apply starch, which involves a rewetting stage prior to further drying. A similar situation exists in mineral coating. These effects are detrimental, as is the mechanical damage that occurs to the fibres in post-surface-chemical treatments such as calendering, supercalendering and corrugating.

There is an inherent danger in just examining the recycling of fibres on their own, since this is a very rare industrial occurrence. The effect of recycling on chemical and mechanical pulps is completely different. Figure

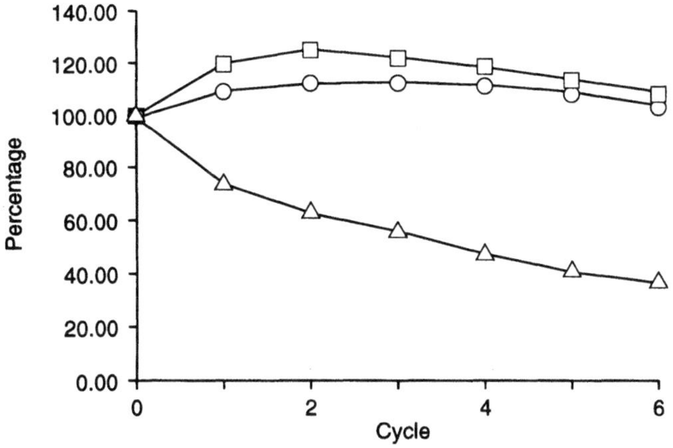

Figure 12.4 Relative changes in paper properties as kraft pulps are recycled. Properties at cycle 0 are taken to be 100%. ○ = freeness; △ = tensile strength; ——— = bulk strength; □ = tear strength.

12.4 illustrates the change in properties for a 'typical' kraft pulp recycled six times. The plot shows the percentage of the original value of the paper made first time or the papermaking stock property. Cycle 0 refers to the first time the fibre is used to make paper, cycle 1 the first time it is recycled and so on. In this example the paper is just reslushed before reconverting into paper. When kraft pulp is recycled, freeness increases, tensile and burst strengths fall and bulk and tear strengths improve. Significantly, the changes are greatest the first time the fibre is dried.

In contrast, mechanical pulp strength properties are essentially unaffected by recycling, and stock freeness reduces. Intermediate pulps, such as heavily modified chemithermomechanical pulp (CTMP) and chemimechanical pulps, have recycling properties somewhere between mechanical and chemical pulps.

The limitation to recycling chemical pulps is related to the reduction in strength, for product reasons. In contrast, the speed at which mechanical pulps drain, a process reason, is their limiting factor. Plotting the data for chemical and mechanical pulps on the same graph, as in Figure 12.5, shows that chemical pulps are stronger for a considerable number of recycles [5]. The consequence of this is that recycled chemical pulps are suitable replacements for mechanical pulps in many applications. This provides a partial explanation for the use of wood-free magazine grades in recycled newsprint – a product that competes head on with newsprint made from mechanical pulp. Similarly, de-inked pulp from office waste, which always contains a mixture of hardwoods and softwoods, is stronger than virgin pulp from hardwoods.

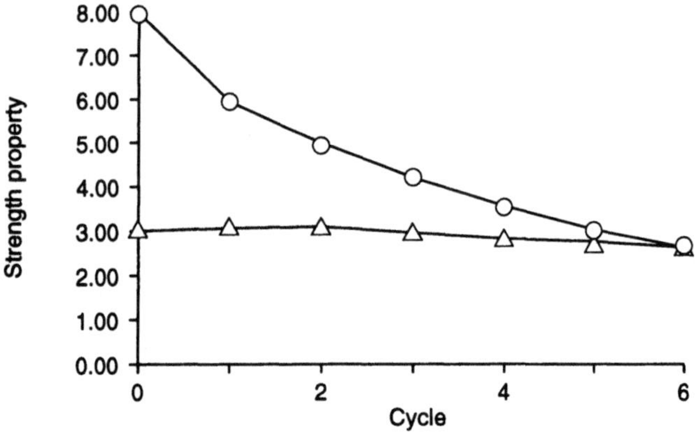

Figure 12.5 A comparison of chemical (○) and mechanical (△) pulps. Source: Guest, D.A., *Pulp and Paper Europe* (January), page 6, 1996.

Conventionally, refining is used to develop the tensile strength of the recycled chemical pulp at the expense of tear strength, bulk and freeness. However, the tensile peak on the energy property curve reduces each time it is recycled, and drainage rate falls. To overcome this problem in corrugating materials, surface applied starch restores the strength properties and is largely responsible for the current high level of recycling in these grades.

The effects of recycling are not all negative. In the case of chemical pulps not all the energy put into refining is lost during drying, with recycled pulp properties being better than those of their unrefined virgin equivalents. For chemical and mechanical pulps the materials released into solution reduce on recycling provided the pH remains below 10. In the case of mechanical pulps where the pH is raised to above this value in de-inking plants a rapid increase in the level of dissolved materials results. The other source of yield loss, non-retained fibre fines, is higher in recycled pulps from either chemical or mechanical processes.

The question frequently posed is the maximum number of times a fibre can be recycled. A much quoted response is five or six. Several workers have tried to model the recycling system to assess the theoretical maximum recycling rate. These tend to show that statistically, at present waste recovery levels, very few fibres reach their fourth recycle. This situation changes rapidly as recovery and re-use rates exceed 50%. The implication of this on waste-paper properties depends on the overall recovery and utilization rates. A number of models exist to calculate the average age distribution of fibres for various recovery rates. By using Cullinan's model [6] and a mixture of 70:30 chemical:mechanical fibres one arrives at the information in Figure 12.6.

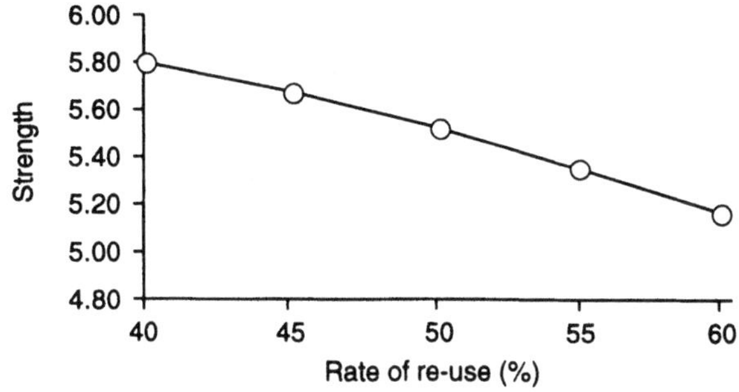

Figure 12.6 The variation of paper strength with rate of re-use.

In the example illustrated in Figure 12.6, the strength (measured as breaking length) of paper made with virgin fibre was about 6.5 km. When about 40% of the fibres have been used before, the breaking length falls to 5.8 km; while, when half the fibres are recycled, it drops below 5.6 km. However, even at recovery rates of as high as 60%, the paper strength, though lower than with virgin fibre, remains at an acceptable level for most applications.

12.2 Recycling surface treated paper products

Surface treated papers are recycled into a range of papers and boards. Preconsumer waste if unprinted will find its way to a mill that manufactures a similar waste-based product, probably with a simple stock preparation system. Once printed, a high-added-value publication paper returns to a de-inking plant, usually making wood-free grades. When it enters the post-consumer stream it mixes with other papers and could end up as a furnish constituent for boxmaking materials, newsprint, tissues and wood-free papers. The effect on the economics and efficiency of the recycling process is grade-dependent.

Additional costs resulting from the inclusion of surface treated papers range widely with the product and installed process. The economic impact is extremely difficult to determine as the effects of surface treatment chemicals, including minerals, are complex. They can be direct, where they cause yield losses and incur extra investment and operational costs for plant and chemicals specifically designed to remove them or ameliorate their effect, or they can have indirect impacts that reduce plant efficiency and make the process more difficult to control.

Although surface applied chemicals have a negative impact on recycling this must not be taken out of context considering the commercial benefits they bring. Surface treatment is an essential process stage in paper manufacturing, adding functionality essential for converting and end use. On recycling, surface applied chemicals affect fibre preparation and papermaking because essentially they behave in one or more of the following ways:

- they remain as discrete particles, for example, coating pigments;
- they stay attached to the fibres and solid particles, for example, cationic polymers;
- they form colloidal particles, for example, coating resins and components;
- they go into solution, for example, starches and carboxymethylcellulose (CMC)

It is also possible for some to move from one state to another, depending on the chemical conditions and temperatures prevailing.

The effects on stock preparation and papermaking are different and will be discussed separately. Stock preparation systems vary considerably in their layout and sophistication and the influence of surface applied chemicals is very much related to the purpose of the plant. In contrast, the influence of surface applied chemicals tends to be independent of the papermaking process, though the extent and costs associated with them vary with the grade of paper manufactured. For this reason, in the following discussion, papermaking is described first by concentrating on the chemicals that find their way through the stock preparation plant.

12.2.1 The influence of separate particles

A major concern to papermakers using waste paper is the presence of mineral coating pigments. In the past the use of chalk pigments in mills using acid sizing gave rise to foaming and deposit problems. The nature of the pigment is no longer an issue as most mills use synthetic sizing chemicals at neutral pH. Minerals are added to most paper stocks used for graphic grades production (newsprint is the major exception) to improve opacity and printing performance. However, coating pigments are finer than the fillers used in wet-end applications and their presence leads to processing and product related difficulties. The primary cause of the problems is their greater specific surface area.

Many chemicals used to control processing carry a positive charge and need to be adsorbed on to the solids to function. A similar situation exists for chemicals used for product performance functions (for example, sizes). Here there is an additional requirement for the solids on which the chemicals adsorb to be retained in the formed web on the first pass through the machine. Additive effectiveness reduces with time, and retention the second or subsequent times around is not as good. Recycled coating pigments adsorb a greater proportion of the process chemicals and this places an added incentive to ensure their retention. However, their first-pass retention is lower than normal wet-end fillers, which increases chemical consumption. The consequence of this is greater chemical costs in the wet end and increased load on the effluent treatment plant. Increased sludge volume and a marginally higher disposal cost result. Also, if the load from the additional solids cannot be dealt with, the discharged effluent loading and cost rise.

The overall water system management strategies and consumption vary widely from mill to mill. Non-retained materials recirculate within closed water circuits and their concentrations increase. In the case of coating pigments this contributes to higher overall stock consistency, a marginal reduction in formation and an increase in energy consumption for pumping. Although first-pass retention of the pigments may be low, their overall retention can be high. This has an impact on sheet strength because, although fillers and coating pigments are both non-bonding, the coating pigments have a greater surface area leading to a greater reduction in sheet tensile strength. This is compensated for by size-press or film-applicator starch addition both in graphics and in corrugated boxmaking material grades. A more intractable problem is the loss in sheet bulk. This can only be compensated for by changing the furnish, refining conditions and pressures applied in the press section and calenders. The effects of coating pigments are not entirely detrimental as the need to purchase filler reduces and both opacity and brightness improve.

Laminates with polymer and metal foil are a special case where the surface treated material stays in one piece and can also remain attached to the fibre. Most mills using waste paper avoid this type of material and rely on the waste merchant to remove it. This is not always possible with mixed waste paper. However, the stock preparation plant usually prevents these materials from passing through to the papermachine. This is dealt with in more detail in Section 12.3 on stock preparation.

12.2.2 The influence of additives that remain attached to the fibres

The effect of these additives runs the whole gamut from negative to positive. Arguably the most damaging are wet strength agents added mainly in the wet end but also in surface treatment stages. The major chemical types are cationic urea formaldehyde, polyamideamine epichlorhydrin, polyamides and melamine formaldehyde resins. These work by forming chemical bonds between the fibres or a polymer matrix that protects the fibres from the effects of water. On repulping they perform their primary function, preventing fibre swelling and the disintegration of the paper. Instead of separating into individual fibres they form bundles.

The stock preparation plant is designed to remove undisintegrated flakes and fibre bundles. However, if these do get into the papermaking circuit and are not removed by the machine screen they cause problems. Inclusion of flakes in the final product detracts from the product value in all but the lowest corrugated board grades. In faster machines the fibre bundles often cause web breaks, with a resultant loss in production. The problem is worst where the waste paper used is from a variety of sources, containing a mixture of wet strength agents. Plants processing preconsumer waste from specific sources can set up the mechanical and chemical conditions to recycle wet strength papers cost-effectively.

Sizing agents are introduced in the wet end and at the surface. Surface applied sizes include alkyl ketene dimer (AKD) wax, alkenylsuccinic-anhydride (ASA), acrylic ester copolymers, styrenemaleicanhydride and styrene–acrylic esters. These form chemical bonds with fibre surfaces and retain some of the sizing effect when recycled, potentially reducing the need for sizing chemicals in publication grades. In the case of AKD sizes the unreacted size hydrolyses on repulping, contributing to sizing problems and causing foaming. This illustrates one of the difficulties of making hard and fast statements about the influence of surface treatment chemicals on recycling. The proportion of AKD reacting varies considerably with formulation, fibre and treatment conditions. Generally, improvements in formulation and application technology are reducing the extent of the problem.

Another category of additives that remain with the fibre or filler are dyestuffs. In mills making coloured papers it is possible to cope with this by using sophisicated colour-control systems. In these cases the impact is neutral. It can become a problem with highly dyed papers that find their way into the post-consumer stream. Mills that specialize in this market area tend to inspect incoming waste paper to remove highly dyed material. Those that re-use their own broke and trim segregate coloured waste and use it for the production of more highly coloured papers.

Strength aids, such as cationic starches, retain some of their activity on recycling. It is not known whether they remain firmly attached to the fibre or detach and then re-attach. Synthetic polymeric dry strength additives, such as cationic polyacrylamides, also have a positive effect when paper is recycled.

12.2.3 The influence of additives that form colloids

Coating latices such as styrene butadiene rubber (SBR), polyvinyl acetate and vinyl acrylates and trace components within formulations form colloids when resuspended in papermaking stocks. Colloidal particles slow down drainage on the machine, but the principal problem is the formation of white sticky deposits on machine components or in the paper. This occurs if the colloid becomes unstable through some change in chemical condition or temperature or through a build up in concentration that suddenly exceeds a critical level.

These deposits, popularly known as white pitch, have a deleterious effect on papermaking economics. Coating mills that recycle their own broke test each new coating additive for compatibility with their wet-end process chemistry and carefully control and monitor the introduction of broke into their system. Problems still occur and mills are sometimes restricted in the range of latices they can use.

For users of purchased-in waste, white pitch is probably only of major concern to those mills that use preconsumer waste for high-grade products. They have limited control of the type of coatings and their components entering the mill and have to take what comes. In the case of packaging grades the white pitch problem is dwarfed by the presence of adhesives from labels, boxes and envelopes. Coating latices, however, contribute to the problem, commonly known as stickies, and the strategies adopted to control their effects are somewhat similar.

Stickies and the problems they cause have existed for nearly as long as waste paper has been reprocessed into paper. Their primary source is the chemicals used downstream of papermaking, including hot melt adhesives for sealing or seaming packaging and pressure sensitives used for self-sealing envelopes and on labels. The problems manifest themselves throughout the machine as deposits in tanks, pipes, on machine clothing and on doctors and drying cylinders. They are also present in the reeled or sheeted product, causing tear-outs on unreeling or sheets to stick together in converting operations.

On chemical analysis stickies are usually complex mixtures of polymers and tackifiers from a variety of sources. Among the chemicals from contact adhesives are SBR, vinyl acrylates, polyisoprenes, polybutadiene and natural rubber. Hot melt components of stickies include polyethylene and ethylvinyl acetate. Some of the ink components found include alkyd resins, polyol acrylates and epoxy acrylates. Many of these polymers are not tacky at the temperatures found in the papermachine water circuits. They absorb tackifiers such as rosin esters and unsaturated fatty acids and complexes formed by the interaction of a range of process additives, including soaps carried over from de-inking plants. It is these tacky complexes that deposit in the wet end of the machine. Deposits in the drying section are caused by polymers reaching their glass transition points; and melting.

The stock preparation system removes most of the adhesive materials. Some components form colloidal dispersions or dissolve completely. It is these that can cause the greatest problem to papermakers. Their concentrations build up in the white water as it recirculates. They can reach a critical coagulation level and precipitate out on parts of the machine or remain with the fibres and become part of the product. The white water is also trapped in the interstices of the forming fabric or within the structure of the press felts. Cleaning showers, if not at the same pH and conductivity, reduce colloidal stability sometimes causing deposition.

There are four practical approaches to control this problem. These are to:

- precipitate or absorb the colloids on the surface of the stock solids;
- disperse the colloids to enhance their stability;
- absorb minerals, usually talc, on the surface of the stickies to reduce their tackiness;
- passivate the surfaces to which stickies adhere.

The precipitation or adsorption on furnish components method relies on the use of highly charged cationic polymers. Its advantage is that it removes the sticky forming materials from the water circulation systems. Polymer additions have to be adjusted carefully to prevent interference with the function of other additives, such as dye fixers and retention aids. Use of cationic additives to fix potentially sticky colloids to fines and minerals in the stock is a comparatively new technique. Exactly how they discriminate between these negatively charged colloids and others is not clear. Possibly they use some form of surface energy difference to take advantage of the hydrphobic nature of the sticky formers. Some of the chemicals are similar to dye fixants and promoters for AKD sizing. They include starch-derived polymers with a branched structure, polyethyleneimines (PEIs), condensation products of organic amides and formaldehyde and poly-DADMAC.

Dispersants can be used to prevent sticky particles from agglomerating and depositing. However, they stay in the system, only being removed in the water in the web at the press section or going to the effluent treatment plant. Their concentration can build up to very high levels, increasing the potential for deposition. Unless carefully controlled, dispersants interfere with the function of other additives and lead to foaming problems. Dispersion shifts the problem to the effluent treatment plant in most mills.

Detackificaion is a traditional method. The use of talc onto which stickies adsorb or adhere, has its origins in pitch control in newsprint mills. It functions successfully on the larger sticky particles that pass through the stock preparation system or form in the wet end. The major disadvantage is that if the dosage and the concentration of the stickies are not matched, surface coverage is incomplete. In some circumstances the adhesive materials are very mobile and the talc adsorbs into the stickie. If this happens in the dryer section, deposition occurs or the product contains sticky particles, causing downstream problems.

Zirconium compounds have also been used to reduce the tackiness of stickies [7]. The mechanism is assumed to be the cross-linking of some of the polymer chains within the deposit, changing its structure and softening point. Whatever the mechanism, effectiveness is improved by trace quantities of dissolved starch.

Passivating surfaces to prevent sticky particle deposition is possible through the application of chemicals that modify surface energy. The prevention of deposition by treating the surfaces on which stickies adhere is to some extent a last resort. This is because the cause of the problem is not tackled and the stickies remain in the system with the potential to cause a problem elsewhere. Deposition tends to be where there are surface energy differences or where a chemical shock occurs. The shower cleaning water must be at the same pH, temperature and conductivity as the stock to prevent agglomeration of stickies in the water in the interstices which leads to their deposition. Some mills use PEI at low concentrations in the shower water, claiming it reduces deposition. It is unclear whether this is a surfactant effect or the PEI adsorbs on the fabric surface, changing its character. Use of sensors on the backwater and shower water to control its make-up and temperature is an interesting possibility, though may not be cost effective.

In practice, mills and their chemical suppliers adopt a holistic approach to controlling stickies. They try to match the cost of the control methods to that of the problem. This is extremely difficult because the problem is constantly changing as downstream and upstream processes alter. A number of new approaches have been suggested using modifications to systems that work elsewhere in papermaking.

Many retention aid systems are now dual applications. Most rely on a low molecular weight, high charge, cationic polymer to remove interfering anionic trash, followed by a high molecular weight anionic or non-ionic polymer. This functions as a conventional flocculant. Another two-polymer system used with high lignin content furnishes is the polyethylene oxide–lignosulphonate combination. The principles may apply to stickies control.

Compozil uses a totally different approach, with cationic starch providing a weakly flocculated matrix, and colloidal silica the controlled flocculation needed for the balance between retention and formation. It is open to speculation whether it is possible to combine the benefits of talc and polymers. Possibly this could be achieved by, for example, treating the talc to adsorb a limited quantity of stickies and then precipitating the particles on fibres.

The only new 'chemicals' on the horizon are enzymes that reduce pitch deposition. Other more specific enzymes may be able to catalyse the 'decomposition' of the natural resins that cause some stickies problems.

The costs of stickies include the capital investment in the plant needed to remove them, the energy consumed by the plant, the use of chemicals for their control and routine cleaning of machine clothing. In addition there are maintenance costs for cleaning deposits from pipes, tanks and doctor blades during routine shuts and lost opportunity costs resulting from machine breaks and the manufacture of non-specification or substandard product. Not many mills calculate these costs systematically, accepting them as a penalty for using lower cost furnishes. In many cases the costs are not easy to access. This is because cleaning plant performs other functions than just removing stickies, and mills do not always determine the cause and cost of web breaks. The proportion attributable to surface treatment chemicals varies. It will be 100% where the mill re-uses its own broke and does not buy in waste paper. It could be lower than 10% for mills making corrugated board materials.

12.2.4 Additives that are soluble

The principal categories are CMC used in coating formulations, and starches used for dry strength or surface improvement applications. They have little affinity with fibres and fillers. Starch is the principal source of dissolved organic material in recycling mills through the use of broke in virtually all non-integrated fine paper mills. The obvious consequences of this are a significant yield loss (some waste furnishes may contain 15% starch). Once in solution, starches are subject to microbiological attack that reduces their molecular weight significantly. This contributes to the built up of slime within the water circuits, particularly where there are 'dead' spots. When slime deposits break away, they frequently cause web breaks and always detract from paper quality. Where anaerobic conditions exist, odour problems result [8].

Mills control this with the addition of biocides in carefully controlled ways. The system can acclimatize to the biocide, which has to be changed frequently to maintain system cleanliness. As well as the cost attached, there are also important environmental considerations when using biocides. Legislation in some countries severely limits the biocides available.

Degraded starch finds its way to the effluent plant and there is an obvious cost attached to its treatment to reduce biological and chemical oxygen demand discharges and the disposal of the biological sludge generated. The alternative of using cationic starch as a replacement for converted or oxidized starch has been examined by a number of mill groups faced with the problem from their own broke. The stark choice is to pay more for the raw material or invest in effluent treatment plant. Most invest in the plant.

12.2.5 Summary of effects of surface applied chemicals on papermaking

Overall the effect of surface applied chemicals on recycling itself are negative and increase the economic and environmental costs of paper-making. These costs are difficult to determine because equipment is very

Table 12.1 Relative recycling costs of surface applied chemicals. Note: − = negative effect; + = positive effect; the more symbols the greater the effect; 0 = no effect

Additive	Woodfree	Mechanical	Tissue	Packaging board
Mineral	− − −	− − −	− − −	−
Cationic polymer	+	0	+	0
Coating latex	− −	−	− −	−
Starch or CMC[a]	− − −	−	− − −	− −

[a]Carboxymethylcellulose

rarely installed just to cope with their presence. Similarly, chemicals added specifically to ensure that others perform their function interact with chemicals originating from processes downstream from papermaking. Papermakers tend to accept the additional costs as an inevitable consequence of using waste paper as a raw material. They attempt to reflect this in the price they are prepared to pay for their waste paper. The overall costs vary with mill practice, waste paper used and the paper grade manufactured.

Subjectively the impacts relative to problems caused by similar materials from other sources or others that have similar effects are shown in Table 12.1 for a number of grades. Coated boards have characteristics similar to wood-free and corrugated packaging board materials. Because the majority of the furnish is in the middle plies, overall the impact of surface applied chemicals will be similar to that of corrugated board materials. By themselves, surface applied chemicals are unlikely to be the determining factor in limiting the recycling rate of paper. The incentive is there to design more benign chemicals and even some that impact positively on recycling.

12.3 Stock preparation and the impact of surface treatment chemicals

The greatest cost arising from the processing of waste paper is the investment needed in-plant to remove the physical and chemical impurities introduced in surface treatment stages, downstream conversion and use. These vary extensively, from the chemicals in adhesives and inks to mechanical objects such as staples and paper clips. These must be separated from the fibres to produce pulp suitable for papermaking. Perfect separation is not necessary as the papermachine can tolerate some level of impurities. However, the impurities incorporated into the product must not detract from its value.

Modern plants use a range of mechanical and chemical techniques, each designed to remove a specific class of impurity. Essentially, they exploit a difference between the fibres and the impurities. These can be chemical, surface chemical or physical, such as size, shape or density. Chemical solubility differences were exploited by Garden State Paper in the USA. They used organic solvents to dissolve adhesives and ink resins. The problems of solvent carryover into the rest of the plant and health and safety caused the mill to revert to conventional technology.

There is a plant in Canada that treats waste paper as a fibrous raw material to be pulped in a similar way to wood and annual crops. The extension of this principle to a solvent pulping process such as Alcel would remove many of the problems met by Garden State. It may, in the long term, provide an interesting solution to the problems arising from surface applied sysnthetic polymers.

Stock treatment plants vary in their complexity. Some are designed to process a specific type of waste paper whereas others are far more flexible. A good example of this are plants that process laminated materials, particularly liquid packaging board. The incentive to use this type of waste is that it generally contains long, strong, bleached fibres. The initial stage in the plant is to treat the packaging board in such a way that the fibres are removed, leaving the laminate with adhesive attached. Exactly how this is achieved is subject to some secrecy [9].

To illustrate the more general effects of surface applied chemicals, de-inking plants, which are the most complex, are used as an example. The function of a de-inking plant is to remove all particles greater than 50 micron or break them down so they are not visible whilst achieving a brightness required for the paper or board product. The economics depend on the costs of energy and chemicals consumed, those arising from the disposal of reject materials and the overall yield on raw material. In tissue mills, minerals detract from the value of the products and there is a supplementary requirement to control their content to a maximum of 5%.

The design of a de-inking system ensures that these impurities are dealt with as well as ink. The impacts of surface applied chemicals can be similar to those described for papermaking or far more subtle because of their impact on the chemistry of de-inking.

De-inking consists of five basic stages. These are:

- the detachment of the ink from the surface of the paper and its suspension in the water phase – this is inevitably influenced by the chemistry of the paper surface and surface applied chemicals;
- the removal of as many ink particles as is practicable or necessary where soluble components of surface applied chemicals perform an important role;
- the dispersion of those remaining particles to reduce their particle size to below 50 microns;
- brightening or bleaching of the stock to the required level, an area influenced by dyes that may be applied at the surface;

- treatment of the water from the process to ensure that it meets discharge consent conditions – this stage parallels some of the problems met in papermaking, though there are some differences important to the efficiency of de-inking and chemical consumption.

In all process stages there is the potential for the dispersed ink to redeposit both on fibres and on fillers. The system also has to cope with the other impurities present. To add to the complexity, the waste paper contains print from a number of processes. These range from conventional offset lithography with oil-based inks, through flexography, where water dispersible inks are used, to laser and xerography where toners are used. The mechanisms for ink attachment to the surface, the resins used and their responses to shear forces vary significantly. Each gives rise to a particular set of processing problems. When considering these factors, it is perhaps not surprising that de-inking plant yields are typically 75%–80% and that they remove as little as 50% of the ink.

12.3.1 Repulping and ink detachment.

Selection and inspection of waste is always a vital part of operating a de-inking plant. Depending on the management philosophy prevailing, either the waste merchant or the mill removes unwanted wastes such as laminated papers and boards, label and envelope waste where glue has been applied and wet strength papers. Where such materials are found every effort is made to trace it to its source to prevent it entering the waste stream.

In the first process stage, mechanical energy and chemicals break the bonds that attach the ink pigment to the paper surface. Surprisingly little work is reported on ink attachment and detachment mechanisms. The bonds broken may be resin-to-fibre, resin-to-mineral, resin-(coating)-to-resin (ink) or resin-to-resin within the ink. In most systems, added sodium hydroxide increases the pH to 10 or higher. This enhances fibre wetting and swelling as well as saponifying resins containing acid groups. An alternative approach is to add a dispersant, some of which are emulsions containing organic solvents. The use of alkaline conditions causes mechanical pulps to yellow. To reverse this effect it is usual to add hydrogen peroxide and sodium silicate, that functions as a pH buffer, dispersant and chelant, to newsprint furnishes. Consequences of operating at this pH are increased release of some surface applied chemicals that can be saponified, lignins and hemicelluloses from the fibres and dispersion of flexographic inks into submicron particles. The latter causes severe problems with removal.

Wet strength additives can be repulped under controlled conditions, but unfortunately these differ for the predominant types. The other problem met is that the wet strength paper entering the mill is random and mixed with other materials. Most mills that cannot cope with wet strength try to exclude it from the furnish. Other additives with a detrimental effect on recycling are sizes. Here the effect of the size is to slow down the rate at which the paper is rewet in the pulper compared with the unsized equivalent. Ink detachment is partly driven by the rate of fibre swelling, so sizing has an influence. The harder the sizing, the worse the effect seen.

After the pulping stage ideally all the coatings and ink particles are detached from the paper. However, there is evidence that some inks and toners fail to detach for reasons that are as yet unknown. Some inks are also attached to overvarnishes (overvarnishing is a specialist surface treatment stage in printing) and are present as large flakes. These can be reduced in size by a specially adapted refiner, known as a deflaker.

The range of ink and other contaminant particle sizes is broad and this renders it impossible to separate them from fibres in a single process stage. The three most widely used separation methods – cleaning, flotation and washing – are most effective in three widely different size ranges. Peak removal occurs at 200 microns, 30 microns and 1 micron, respectively. The early de-inking plants used either a flotation or a washing strategy, with the former preferred in Europe and the latter in North America. Nowadays, plants throughout the world incorporate both washing and flotation in some form or another.

During stock preparation the pumps cause impurity particles to break down. The cleaning strategy always has the separation plant most effective on large particles first and that most effective for smaller particles last. Comparatively simple hydraulic screens remove large particles and bundles of fibres held together by wet strength agents after the pulping stage.

Other mechanical separation stages through the plant rely on density differences between the paper fibres, fillers and impurities to effect removal. Centrifugal cleaners are not very effective separators and reject usable fibre with impurities. The rejects pass through secondary and tertiary stages to recover as much fibre as possible and concentrate impurities into the reject stream. Accepts from these stages feed into the preceding stage or to the accepts from the primary stage. Stock cleanliness and the capacity of the plant are critical determinants of the route selected.

12.3.2 De-inking and de-ashing

Passing the stock through low-density cleaners prior to flotation removes flakes of undisintegrated surface applied chemicals along with large ink particles. Flotation systems have developed considerably since their early adaptation from the plant used for mineral separation. Essentially, air is bubbled through stock as it passes through an agitated cell. The air bubbles attract the ink particles, which rise to the surface where they are trapped in the foam layer. The foam containing the separated ink particles is skimmed off or passes over a weir.

A key stage is to change the surface chemistry of the ink particles to increase the attraction to and the strength of the interaction with air bubbles [10]. Ideally the bubbles and the inks should be similarly hydrophobic. Traditionally, fatty acid soaps, added at the pulping stage or in a mixing chest, were used, though this approach is under challenge from a range of synthetic surfactants. Some of these have been modified from those used in the laundry industry. The hydrophobicity is developed by converting the acids to calcium salts that precipitate on the surface of the ink particles.

Process variables that influence the effectiveness and efficiency of de-inking include stock consistency, throughput, pH, the speed and hence shear forces caused by agitation, the bubble size distribution and the air:stock ratio. Mixing and shear forces affect the number of collisions that occur between the ink particles and air bubbles. The effectiveness of a set condition on ink particle flotation has different results depending on ink particle size. Two factors are important: for smaller particles the energy of collision must be sufficient to overcome the barriers to ensure adhesion; with larger particles there must be sufficient momentum to keep the air with attached ink rising to the surface. Decreasing zeta potential causes agglomeration of smaller particles to a size suitable for flotation. Soluble materials in coatings with a negative charge in water change the zeta potential and consequently the ink removal efficiency. Surface active materials also influence air-bubble size. Since there is a direct relationship between bubble size and ink removal efficiency, surface applied chemicals play an important secondary role in de-inking.

The influence of surface treatments, particularly pigmented coatings, on ink removal efficiency has been studied at laboratory and mill scale. Experience indicates that mineral coatings are beneficial, though the evidence is more qualitative than quantitative. (Oddly, the coating has to be dried for the effects to be measurable.) A number of workers have unsuccessfully attempted to isolate the reason for this in the quest to discover a new de-inking aid. The reported efficiency improvements may be a result of the way the experiments were undertaken and the measurement techniques used. It is only recently that direct measurements on ink removal efficiencies have been made with in most cases measurements inferred from stock brightness. This is influenced by the initial brightness of the fibres, the mineral concentration, the mass of ink and just as importantly the ink particle size distribution. The work also was not designed to examine where in the de-inking process the positive effects originated nor the influence of yield on the results.

Some clues may emerge from work looking at the nucleation and growth of the calcium soaps so important for attracting ink particles and in the flotation stage itself. A key stage in flotation is to change the surface chemistry of the ink particles to increase the attraction to and the strength of the interaction with the air bubbles. Ideally the bubbles and the inks should be similarly hydrophobic. The traditional and more common approach is to add fatty acid at the pulping stage or in a mixing chest just prior to flotation. Calcium ions interact with the saponified soap forming hydrophobic agglomerates that coat the ink particles' surfaces. The exact mechanism is the subject of some controversy, though two aspects are important: the time taken for nucleation to start and the rate at which agglomerates grow [11].

The speed of agglomeration increases with greater agitation and increasing calcium and soap concentrations. Silicates and pectic acids reduce the time before the onset of agglomeration and its rate, and CMC, starch and non-ionic surfactants decrease agglomeration rate by improving colloidal stability. The structures of the agglomerates change with time, moving from an open structure to a more dense final structure. Soluble components of surface applied chemicals thus play an important role.

Tissue grades require low levels of filler and coating pigment for product reasons. Recently a systematic study on flotation as a means to remove filler and coatings has started at Helsinki University of Technology. This work will separate the effect of the filler itself from its interactions with ink in a flotation cell.

Stickies are removed along with ink in de-inking mills at both the washing and at the flotation stages. The flotation system is designed for removing ink, and stickies removal is coincidental. Work on modifying conditions has improved removal efficiencies for stickies, though there are no reports of these stages being installed primarily for this purpose.

Flotation is not very effective at removing smaller particles; as a rule of thumb it is most effective at removing particles with diameters seven times smaller than those of the air bubbles. For smaller particles, washing is the only viable separation technique. This is a simple technique in which the stock is passed over a fine mesh. Smaller particles, including the ink, pass through the mesh with the majority of the water and are rejected from the system. The mat formed is then diluted before repeating the process again. The number of stages used is an economic matter, but can be calculated from a theoretical standpoint once separation efficiences are known. In practice once a mat of fibres is formed separation efficiency is limited by the pore sizes within the mat and not by the size of the mesh used. Fillers and minerals reduce pore size and de-inking efficiency, though there are no reported studies on how big this effect is.

To reduce water usage in washing it passes countercurrently through the plant, fresh water only being used for dilution prior to the last stage. A number of washing devices are used, varying in the inlet and outlet consistencies and hence in the water volume needed. Typical examples are

Sidehill screens, deckers and screw presses. Since particle size is important, the presence of dispersants from surface treated papers has a minor impact.

New approaches to de-ashing, based around fractionation, are finding their way into mills, with the first installation in the United Kingdom [12].

12.3.3 Dispersion

There are still ink particles visible to the naked eye that detract from product appearance. In most recent de-inking plant installations, stock is thickened and then dispersed. Essentially, this subjects the fibres and inks to a temperature in the region of 90°C–120°C and mechanical action. This softens the polymers holding the ink agglomerates together and reduces their size to below 50 microns. Stock is thickened to 30% consistency. The factors affecting thickening are similar to washing. Colloidal materials present from surface treated wastes will have a negative impact on water removal and possibly on the effectiveness of dispersion.

12.3.4 Colour stripping

Where post-consumer waste is a major part of the furnish it is difficult to exclude dyed papers. Although the majority of the dye is added to the stock in the wet end some may be applied via a surface treatment stage. Some chromophores in the dyes are not oxidized by hydrogen peroxide under the conditions typically used in de-inking plants of 90°C at 30% consistency for up to an hour. It is usual to include a second stage involving reductive treatment with sodium dithionite. Evidence from the USA suggests that this approach is not suitable as waste recycling levels increase. Since many de-inked pulps and waste-paper-containing papers are sold on their environmental credentials, chlorine-containing oxidizing agents cannot be used. A new approach is treatment under pressure with oxygen and hydrogen peroxide. Most recently, plants using ozone bleaching have been installed in the USA.

12.3.5 Water treatment

The water circuits within the de-inking plant contain a mixture of suspended solids, including fibre fragments and fines, minerals and ink pigments. They also contain colloidal materials and a variety of dissolved organic materials from the waste and the chemicals added to the system. Cleaning this up prior to circulation or discharge is a complex operation. Solids are flocculated using cationic high molecular weight polymers and separated using dissolved air flotation. Secondary biological treatments remove the dissolved organic materials and reduce the biological and chemical oxygen demand levels.

The water recirculated has a significant impact on brightness after de-inking. Experiments using ultrafiltration (UF) to remove the colloidal material demonstrated a significant reduction in the cationic demand of the aqueous phase. A combination of microflotation and UF halves chemical oxygen demand from 4 g l^{-1} to 2 g l^{-1}. UF permeate use in pulping and flotation in place of raw dilution water led to a 2%–3% ISO brightness improvement.

The solids are further treated to reduce their moisture content in screw and belt presses. Traditionally, the sludge was disposed of by landfillng. This option is becoming restricted either for legislative or for cost reasons. More mills are turning to incineration with energy recovery as an alternative. Minerals from surface applied coatings have a negative impact by increasing sludge volume and reducing its calorific value.

12.4 Summary of the effects of surface treated papers on recycling

Papers are surface treated to provide functional properties. These include improved printing performance, optical properties, resistance to water and humidity and enhanced surface and bulk strength. The range of chemicals and uses makes it difficult to draw firm conclusions about whether the benefits they confer during conversion and in use outweigh their negative impacts on recycling. Some of the major impacts of surface applied chemicals are summarized in Table 12.2.

What also adds to the difficulty is that many chemicals can be applied either in the wet end or on the surface of the formed web. With modern surface application technology and the inherent difficulties of controlling the wet-end surface application is starting to gain market acceptance and share. There is no reason to expect that synthetic size application at the surface will be any more detrimental to recycling than addition at the wet end. Indeed, because the application is more selective, less chemical is used and the overall impact is likely to be lower.

Coated papers have far greater impact than uncoated papers on the costs associated with recycling. In common with most products, coated papers are not designed to be recycled, nor were they designed with any particular waste management strategy in mind. It is not too surprising that they give rise to problems when recycled. These problems will be relatively more costly where they return as high grade waste paper in the form of trim and waste from converting processes. These waste papers are almost exclusively used to manufacture high-grade products. The potential problems arising from the use of coated waste in high-grade products are white pitch and the low retention of the coating pigments. The latter increases the disposal

Table 12.2 The principal effects of surface application on recyclability

Chemical	Effect on recycling
Water	Reduces pulp strength – a result of wetting and redrying
Mineral pigments	Improve brightness and reduce need to purchase filler
	Increase yield losses
	Interfere with the function of wet-end additives
	Increase wet-end chemical costs
	Increase effluent treatment costs
	Substantially increase costs for landfilling
	Reduce calorific value of sludges when incinerated.
Native/anionic starches	Increase yield losses by a minimum of 7.5%
	Interfere with the function of wet-end additives
	Increase wet-end chemical costs
	Require biocides to reduce effects of microbiological degradation.
	Increase effluent treatment and disposal costs
	Modify performance of calcium soaps in de-inking.
Cationic starch	Minor yield losses
	Positive impact on paper and board strength
Synthetic latices	Cause white pitch problems and contribute to stickies
	Increase costs for cleaning machinery, felts and fabrics
	Reduce productivity and cause spoiled product
	Modify nature of bond with ink vehicle and ink detachment mechanism
Synthetic sizes	Increase energy needed to disperse fibres
	Unreacted alkyl ketene dimer contributes to foaming
	Reduce the need for sizing chemicals
Wet strength chemicals	Increase complexity of plant needed to disperse fibres
	Cause machine breaks
Dyes	Restrict recycling to coloured grades unless bleaching plant installed
Laminates	Restricted to plants designed to process them
	Add to the complexity of processing
	Yields are usually low and waste disposal costs high
Carboxymethylcellulose	Makes minor contribution to yield losses
	Modifies formation of calcium soaps in de-inking plants

costs of sludge to landfill, whether or not it is incinerated. On the positive side, a proportion of the mineral is included in the base sheet, reducing the overall need to extract and process mineral.

After reaching the consumer, there are two collection and recycling routes. If the consumer segregates the coated waste paper by returning to a collection point it is likely to find its way to a newsprint de-inking plant. The problems listed above still pertain but are significantly outweighed by the benefits. These include the stronger fibre furnish and the enhanced brightness from the minerals present. Their partial inclusion in the product enhances surface properties and printability.

Unsegregated waste will eventually find its way into the furnish of low-grade board, testliner and fluting. Here the white pitch problem is dwarfed by that caused by adhesives from labels and packaging. The soluble polymers in the coating are relatively minor contributors to the dissolved organic loads.

However, the functions and performance of coated and uncoated papers are different. They find use in different value networks and a direct comparison of their recyclability is unfair. If judgements are to be made it is essential that the products are destined for the same market and perform equally well. The comparison also must consider the total impact on the environment and not just be focused on the solid-waste aspects.

Coated board wastes contain a mixture of chemical and mechanical fibres and are used in newsprint or packaging product manufacture. Here the impact of the coating is far less serious. Surface treatment of wood-free papers with starch has a detrimental effect caused by yield losses and the extra process chemical costs. This can be reduced by using cationic starch where there is an economic trade-off between the extra costs of cationizing the starch and the reduced costs of recycling.

In the case of corrugated packaging materials they are usually used for the manufacture of similar products. Some material is sourced from converting plants or within the distribution chain. More is from the consumer. Corrugated boxes contain up to 15% starch added in the size press and during corrugating. Although users of OCC and mixed waste frequently complain about the declining standard of waste paper they seem to cope with these levels of soluble starch. They also compete very successfully with producers of virgin materials, as evidenced by their greater market share. In the longer term there must be a concern that as less virgin-based product finds its way into lower grade wastes the increasing use of minerals in coatings will cause problems for which surface applied starch cannot compensate.

It is impossible with the present state of knowledge to reach any conclusions on the levels of surface treated paper and boards that would result in the generation of a ceiling to paper recycling. Judgementally, this is much more likely to result from the presence of other chemicals and materials added during conversion and use. The use of non-renewable energy to extract and refine minerals may have a greater environmental impact on the manufacture of coated products than the additional impacts that occur during its recycling. However, mineral suppliers need to pay some attention to the potential uses for sludge from de-inking plants, both in the wet state and after incineration, to further reduce the environmental impacts. It may be possible through dispersion or separation technology to develop products with the potential to become a raw material for another process. The environmental and economic benefits of this need careful attention.

Future developments in surface applied chemicals have to consider the recyclability issue and impacts at all points in the product value network. There may be lessons to be learnt from other areas where chemicals cause problems, for example, adhesives. Repulpable or recyclable adhesives work on a number of operating principles. In their simplest form, in splicing tapes for in-mill use, the adhesive is soluble or dispersible. The more sophisticated approach is to modify the adhesive formulation to resist the mechanical forces in repulping. Where this is successful the adhesive remains intact or stays in pieces large enough to be removed by the coarse screens. An alternative is to modify the adhesive's density to facilitate its removal in high-density or low-density cleaners. The metal powders, high-density fillers and modified polymers proposed have yet to find commercial application.

This possibly sums up one of the major dilemmas. A recyclable product has to perform as well as the one it replaced during application and end use. Chemical manufacturers should not be discouraged in their efforts to produce recyclable products. They must, however, understand more closely the impacts of their products on the economic and environmental costs of recycling.

References

1. Ervasti, I. (1995) Recycling. '95 Proceedings, October, 1995. Paper Industry Technical Association, Bury, p. 159.
2. Pulp and Paper International (1995) Annual review. July, p. 27.
3. Bruce Darlington, W. (1995) Wastepaper '95 Proceedings, June, 1995. Pira International, Leatherhead, paper 22.
4. Howard, R.C. (1990) Annual PPRIC Meeting, p. B337.
5. Guest, D.A. (1996) *Pulp and Paper Europe* (January), 13.
6. Cullinan, H.T. Australian Pulp and Paper Industry Technical Association, **45** (1) 6.
7. Mann, S. (1995) Wastepaper '95 Proceedings, June, 1995. Pira International, Leatherhead, paper 13.
8. Gudlauski, D.G. (1996) *Pulp and Paper* (March), 161.
9. Huston, J. (1995) Wastepaper '95 Proceedings, June, 1995. Pira International, Leatherhead, paper 21.
10. Guest, D.A. (1993) *Products of Papermaking* (September), 1141.
11. Guest, D.A. (1996) *Pulp and Paper Europe* (January), 29.
12. O'Brian, H. (1996) Pulp and Paper International (March), 46.

Index

Page numbers in **bold** refer to figures and page numbers in *italic* refer to tables.

GPSR Compliance

The European Union's (EU) General Product Safety Regulation (GPSR) is a set of rules that requires consumer products to be safe and our obligations to ensure this.

If you have any concerns about our products, you can contact us on ProductSafety@springernature.com

In case Publisher is established outside the EU, the EU authorized representative is:

Springer Nature Customer Service Center GmbH
Europaplatz 3
69115 Heidelberg, Germany

Batch number: 10755332

Printed by Printforce, the Netherlands